Topology and Physics of Circular DNA

Alexander Vologodskii
Institute of Molecular Genetics
Russian Academy of Sciences
Moscow, Russia

Translated into English by D. Agratchev

CRC Press
Taylor & Francis Group
Boca Raton London New York

CRC Press is an imprint of the
Taylor & Francis Group, an **informa** business

First published 1992 by CRC Press
Taylor & Francis Group
6000 Broken Sound Parkway NW, Suite 300
Boca Raton, FL 33487-2742

Reissued 2018 by CRC Press

A Library of Congress record exists under LC control number: 92009528

Publisher's Note
The publisher has gone to great lengths to ensure the quality of this reprint but points out that some imperfections in the original copies may be apparent.

Disclaimer
The publisher has made every effort to trace copyright holders and welcomes correspondence from those they have been unable to contact.

ISBN 13: 978-1-138-10505-8 (hbk)
ISBN 13: 978-1-138-56281-3 (pbk)
ISBN 13: 978-1-315-12130-7 (ebk)

Visit the Taylor & Francis Web site at http://www.taylorandfrancis.com and the CRC Press Web site at http://www.crcpress.com

PREFACE

Circular DNA holds a special place among biological macromolecules. Its unique physical properties are entirely generated by a specific topology which must be retained through every conformational change. The topological constraints imposed on the conformational range of circular DNA molecules make for a very different set of properties as compared with linear DNA.

Biochemists and molecular biologists have shown a growing interest in the properties of circular DNA in the last decade. There are two major reasons for that. On the one hand, it has gradually transpired that circular DNA is the key form for the functioning of DNA within the cell. On the other hand, there is an increasing body of data to the effect that the topologically defined properties of circular DNA play a critical part in biological processes. The above facts are the rationale for writing this book.

Every effort has been made to minimize the requirement of advance special knowledge and thus to make the book accessible to a wide range of readers. To this end, an introductory chapter has been written, detailing those properties of linear DNA that are referred to in the subsequent text.

There is one more reason why circular DNA is particularly interesting; it is an excellent illustration of topology's impact upon the properties of polymeric systems. In the instance of circular DNA, topologically defined properties are especially salient. Even now, these properties are quite well researched, in both theory and experiment. The rapid development of molecular biological techniques creates unique opportunities for further experimental studies, wherein DNA molecules can be subjected to the finest analysis. In a variety of cases, circular DNA may prove to be the most appropriate material for the study of topological effects in diverse polymeric chains.

Chapter 1 is an introduction to the general properties of DNA. The main concepts pertaining to and general properties of circular DNA are presented in Chapter 2 . Familiarity with the matters treated in this chapter is essential to the understanding of the following chapters, which are relatively independent from one another. This structural organization of the book reflects itself in the attached bibliography. Most of the references in Chapter 1 are to monographs and reviews, whereas the other chapters primarily cite original works.

Since studies of circular DNA have given rise to some elegant and potent experimental approaches, the reader will find descriptions of certain unique methods alongside the explanation of theoretical concepts and experimental results.

Circular DNA has proved to be not only a fascinating object of research, but a powerful tool in the study of general properties of DNA. Chapter 5 deals with the use of circular DNA in establishing various characteristics of the double helix.

The author gives little or no attention to the biological role of the various properties of circular DNA, the matter being beyond the scope of this monograph. It is, however, the author's remise that a clear understanding of circular DNA's physical properties is essential to any analysis of the ways in which topology influences the biological functions of DNA.

The author would like to acknowledge the people who have, in different ways, stimulated the writing of this book. They are my colleagues who have worked with me on many of the problems treated   below: M. D. Frank-Kamenetskii,   Yu. S. Lazurkin, A. V. Lukashin, V. V. Anshelevich, V. I. Lyamichev, and S. M. Mirkin. Thanks are also due to my wife M. Yu. Pokrovskaya for her help in manuscript preparation, V. I. Lyamichev and B. R. Amirikian, who read the manuscript and did not withhold their comments and criticisms, and E. M. Shekhtman for assistance in preparation of the figures.

A. V. Vologodskii

THE AUTHOR

Alexander Vologodskii is a Senior Scientist of the Institute of Molecular Genetics of the Russian Academy of Science, Moscow, Russia.

Dr. Vologodskii graduated in 1972 with a degree in molecular biophysics from Moscow Physical-Technical Institute, where in 1975 he obtained his Ph.D. In 1985 he received his D.Sc. in physics and mathematics from Moscow University.

Dr. Vologodskii has presented over 20 invited lectures at national meetings and numerous guest lectures at universities and institutes. Since 1989, after liberalization in his country, he has been invited to various international meetings. He spent 1 year at the University of California, Berkeley, as a Visiting Professor. He has published more than 60 research papers. Dr. Vologodskii is well known for his work on topological properties of circular polymer chains, DNA supercoiling, and conformational transitions in DNA. His current major research interests include conformational properties of supercoiled DNA and knots and catenanes of DNA.

TABLE OF CONTENTS

INTRODUCTION

In 1963, R. Dulbecco and M. Vogt found DNA to exist in closed circular form in certain viruses. In this form, the two single strands that make up the double helix are closed in upon themselves. The complementary strands are linked, with a fairly high linking number. At that time, one could hardly suppose that this would prove to be the prevalent form in which DNA functions within the cell. As time went by, however, circular DNAs were discovered in an increasing number of organisms. By the mid-1970s, the giant DNA molecules of higher organisms were found to consist of multiple loops, each of which was analogous in its properties to circular DNA. Just then, enzymes were discovered that controlled the level of supercoiling, a key property of this DNA form. Gradually, it became quite clear that closed circular DNA and the supercoiling phenomenon associated with it had an extremely important role in living organisms.

In physical terms, the chief distinctive feature of closed circular molecules, as compared with linear DNA, is that once the complementary strands lock into a circle, the molecule assumes a topological state that cannot be altered by any conformational rearrangements short of breaking the strands. Because of this topological limitation, many of circular DNA's physical properties are drastically different from those of linear molecules.

The study of the physical properties of circular DNA began in the same year, 1963, at J. Vinograd's laboratory. It was thanks to this group's efforts, plus a major contribution by J. Wang and some other researchers, that a considerable body of knowledge about these unusual properties was accumulated by the late 1970s. Among the most important results, one can cite reliable methods for determining the linking number difference in circular DNA and the energy characteristics of supercoiling. Furthermore, it was found that the double helix is more flexible in the supercoiled state and more prone to fluctuational disturbances. A pivotal method of gel electrophoresis has been developed, whereby DNA molecules that only differ in the linking number of complementary strands in the double helix can be separated mobility-wise. Finally, F. B. Fuller came up with a mathematical description of supercoiling based on the ribbon theory.

The late 1970s saw a new powerful upsurge in the studies of circular DNA. At that time, the wide-spread use of genetic engineering and DNA sequencing techniques raised experimental research to a qualitatively new level. Major results did not take long to follow. Negative supercoiling was found to cause entirely new structures, quite different from the conventional double helix, to arise in circular DNA. These structures, whose existence is disadvantageous to linear DNA under normal conditions, include the cruciform, the left-handed helical Z-form, and the H-form, which is based on a triple helix. The discovery of these structures naturally raised questions as to their possible biological role, which is currently being looked into with considerable intensity.

It would be no exaggeration to say that the studies of circular DNA, its properties and related issues, are currently among the most intensive areas of molecular biophysics. As in any rapidly developing field, there are many questions that remain to be answered. The purpose of this book is to sum up the work accomplished in this fascinating field over the last three decades and to attempt some generalizations based on the results.

Chapter 1

BASIC PHYSICAL PROPERTIES OF DNA

I. DNA STRUCTURE

This section deals with the principal structural characteristics of DNA. We should point out at once that atomic structure is beyond the scope of this book. We are going to look at structure primarily as it relates to the analysis of circular DNA and its properties. Therefore, many important issues will be overlooked or barely touched upon. For a more comprehensive overview of DNA structure, the reader is referred to the monographs and surveys cited in this chapter.

A. Single-Stranded DNA

Single-Stranded DNA consists of a sugar-phosphate polymeric backbone and lateral nitrous bases attached to it. The chemical structure of single-stranded DNA is presented in Figure 1. The backbone's repetitive unit (link) includes six skeletal bonds. In each chain link one of four bases, adenine (A), guanine (G), cytosine (C), or thymine (T), is attached to the C1' atom of the sugar ring. These bases (see Figure 2) belong to two different chemical groups; adenine and guanine are purine bases, while cytosine and thymine are pyrimidine bases. The base sequence, unlike the sugar-phosphate chain, is irregular, for it contains the coded genetic text carried by the DNA molecule. As can be seen in Figure 1, the backbone has no symmetry relative to the movement along it one way or the other; the structure of the backbone defines the chain direction. One end of the chain is called the 3' end (lower end in Figure 1) and the other is called the 5' end (upper end in Figure 1). These appellations have to do with the numbers of carbon atoms in deoxyribose. To define the chemical structure of a single-stranded DNA, it is sufficient to specify the base sequence. By a generally accepted convention, the sequence is always written from the 5' end towards the 3' end. Rotation may occur around each single bond of the sugar-phosphate backbone. Naturally, this rotation is not free because a number of rotation angles (called dihedral angles) entail too close contacts between atoms adjacent to the principal chain. Still, the permissible dihedral angle ranges are large enough. (For more detailed and rigorous presentation of this question, the reader is referred to the Cantor and Schimmel textbook.[1]) Therefore, the sugar-phosphate backbone is highly flexible; even one repeating unit can assume a very large number of conformations. Without causing any major tensions in the structure, one can adapt the sugar-phosphate backbone to a variety of conformational requirements. This property is

1

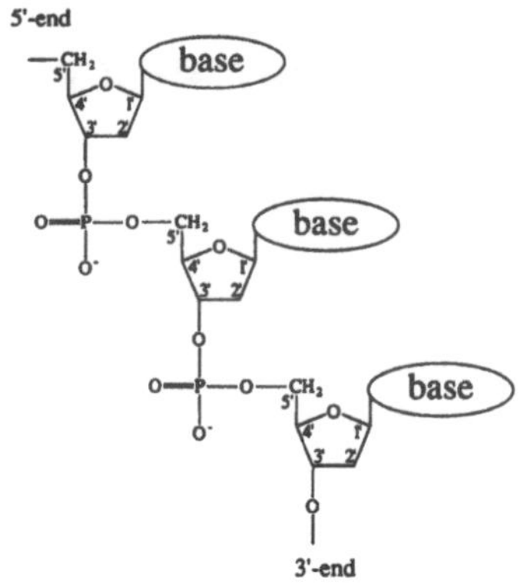

FIGURE 1. Chemical structure of single-stranded DNA.

responsible for the polymorphism of helical structures that one observes in double-stranded DNA. Note that the size of a repeating unit in a conformation stretched to the limit is about 0.7 nm.

A part from rotation around backbone bonds, there is also the very important rotation around the so-called glycosidic bond between the sugar and base. Permissible rotation angles around the glycosidic bond largely fall within two ranges with almost 180° in between. The nucleotide link conformations corresponding to these two ranges are designated as *syn* and *anti*. Figure 3 shows the *syn* and *anti* conformations of purine and pyrimidine. A large body of calculations and experimental data obtained with monomers demonstrates that the two conformations are roughly equivalent energy-wise for purines, whereas the *anti* conformation is the referred one for pyrimidines.[1]

Finally, another degree of freedom of the polynucleotide chain is associated with the conformation of deoxyribose. This ring's conformation is not flat. As a rule, four atoms lie on a plane, while the fifth one, either C2′ or C3′, is situated above or below the plane. Depending on whether or not this atom is shifted off the plane in the same direction as C5′, the conformation is denoted as *endo* or *exo*, respectively. Normally, C2′-*endo* and C3′-*endo* conformations occur in regular nucleic acid structures.

The sugar-phosphate chain carries a single negative charge per phosphate group. This has an appreciable effect on the conformational properties of DNA. Various consequences of DNA's polyelectrolyte nature will be discussed throughout this book.

B. The Double Helix

Single-stranded DNA is a relatively rare thing under natural circumstances. One can say that DNA's regular form is a double helix consisting of two antiparallel single strands. The discovery of the double helix by Watson and Crick in 1953 is regarded as the birthdate of molecular biology. More than

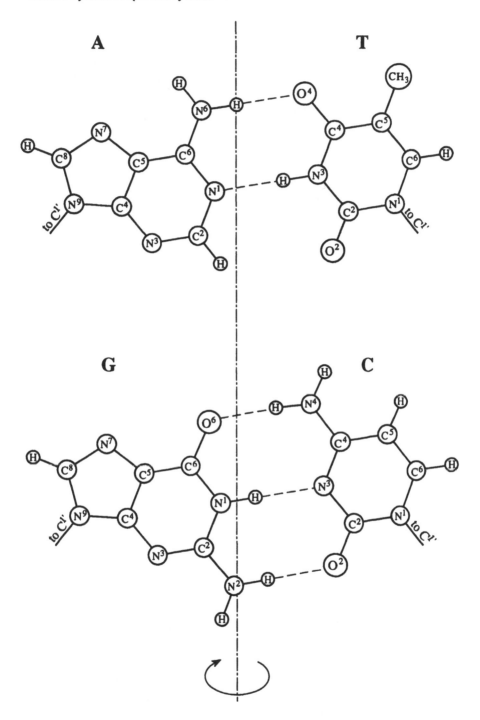

FIGURE 2. Complementary base pairs. On turning around the pseudosymmetry axis that passes between pairs, the conds N9-C1' and N1-C1' trade places.

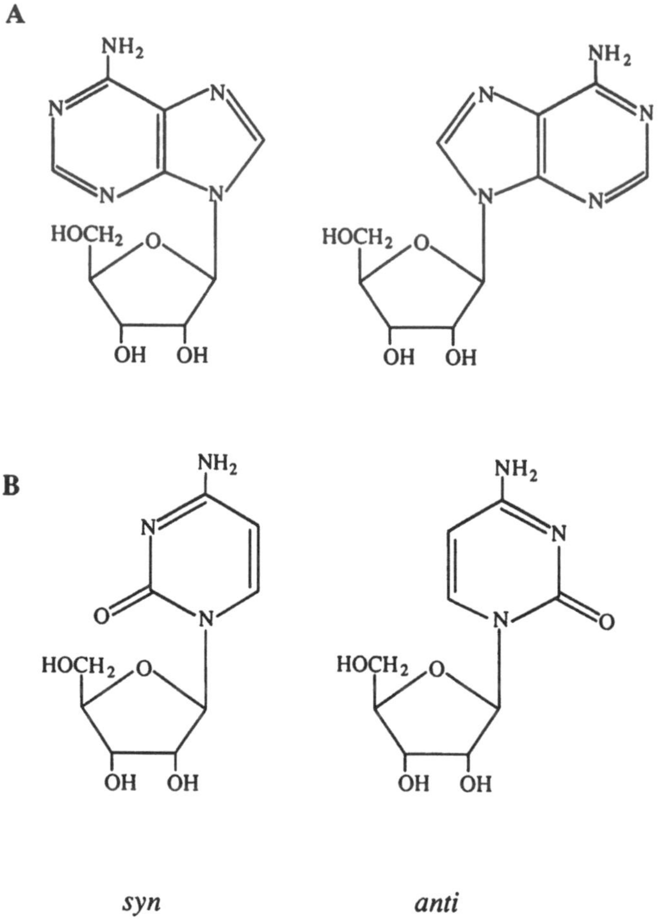

A

B

syn   anti

FIGURE 3. *Anti* and *syn* conformations of purine (A) and pyrimidine (B).

the structure of the genetic information carrier was discovered at that time. Even a fleeting glance at this structure makes it clear that it contains the mechanism of a fundamental genetic process, *viz.*, the duplication of genetic information. No biological structure decoded before or after that momentous discovery proved nearly so revealing of the molecular functioning mechanisms.

A three-dimensional model of the double helix (B-form) is shown in Figure 4.[3] The bases in this structure are inside the helix, while sugar-phosphate chains are outside, running in opposite directions. Thus, the double helix, unlike a single strand, does not have a structurally defined direction. The bases, one from each strand, form pairs linked by hydrogen bonds. Significantly, only two types of pairs, AT and GC, are possible in a regular double helix. These pairs have the same geometry with regard to their bonds to the sugars. Besides,

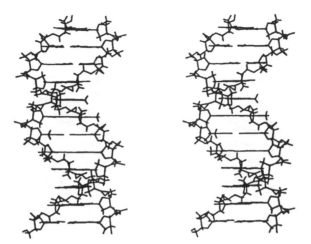

FIGURE 4. Skeleton model of B-DNA double helix. (From Schlick, T., et al., *Theoretical Biochemistry and Molecular Physics, Volume 1: DNA,* Adenine Press, Schenectady, NY, 1980. With permission.).

their symmetry lane passes through the helix axis (see Figure 2). Therefore, they are readily inscribed in the regular double helix structure. These pairs are called Watson-Crick, or complementary, pairs because the two bases complement each other in forming the characteristic pair geometry. The requirement that the bases within a pair be complementary is the reason why the sequence of one strand fully defines the sequence of the other strand.

The base pairs are situated inside the helix and cling tightly to each other, forming a pile. That is why they have almost no contact with water molecules, and their accessibility to chemical reagents is severely limited. This fact is widely used in studying various disturbances of the regular DNA structure.

The structure of the double helix may change depending on the ambient solvent. In aqueous solutions under near-physiological salt conditions, the double helix assumes a structure called the B-form (see Figure 4). In the B-form, the DNA helix is right-handed; the base pairs (bp) are practically flat and lie perpendicular to the helix axis. The helix axis passes close to the center of each pair. Sugar is in the C2′-*endo* conformation, while all the bases hold an *anti* conformation in relation to the glycosidic bond. The double helix has two grooves, one of which lies between the two strands' sugar C1′ atoms and is called the glycosidic or minor groove, while the other lies opposite the first one on the other side of the helix and is called the nonglycosidic or major groove. The pitch of the double helix is an important parameter of DNA. The average period of the B-form is 10.5 bp per turn. The pitch of the B-form double helix may change somewhat with changing ambient conditions (see Chapter 2, Section II.B). The characteristic geometric parameters of B-form DNA are as follows: the helix diameter is 2.0 nm, and the distance between

A –DNA

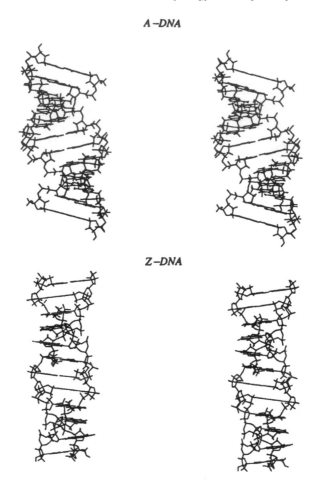

Z–DNA

FIGURE 5.   Skeleton models of A-DNA and Z-DNA. (From Schlick, T., et al., *Theoretical Biochemistry and Molecular Physics, Volume 1: DNA,* Adenine Press, Schenectady, NY, 1980. With permission.).

pairs along the helix axis is 0.34 nm. It has transpired in recent years that fine geometric details of the B-DNA structure depend on the local sequence,[2] but the parametric changes involved are fairly small.

C. Alternative Forms of the Double Helix

Two double helix structures other than the B-form are known at present. The first of these is called the A-form and occurs in DNA fibers at a low relative humidity or in solutions with a low solvent activity. Under normal conditions, the RNA double helix is in this form as its additional sugar hydroxyl group prevents RNA's sugar-phosphate chain fitting within a B-helix. Like the B-form, this is a right-handed helical structure (see Figure 5).

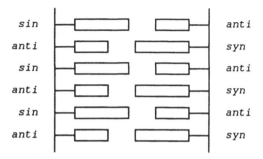

FIGURE 6. Alteration of *anti* and *syn* base conformations in a regular Z-helix. The large rectangles correspond to purine bases and the small ones to pyrimidine bases.[4]

According to fiber diffraction analysis, the helix pitch in this case is 11 bp per turn. No direct data are available as to the helix pitch in solution. As in the B-form, the base pairs are almost flat in A-DNA, though here they are tilted by 20° in relation to the plane perpendicular to the helix axis and are removed from this axis. The distance between bases, measured along the helix axis, is 0.26 nm in A-DNA. Sugar is in the 3'-*endo* conformation. The fact that the A-form differs from the B-form in sugar conformation points to the existence of an energy barrier separating the two forms. The orientation of all bases relative to the sugar corresponds to the *anti* conformation.[1]

For a long time, the B- and A-forms, or rather the B- and A-families of forms, were believed to cover the entire conformational range of the double helix realizable in nature. Then in 1979, quite unexpectedly, A. Rich and co-workers found the structure of the self-complementary hexanucleotide d(CG)$_3$ in their crystals to be a left-handed double helix. The sugar-phosphate chain in that structure did not run in a smooth line as in the A- or B-forms, but zigzagged, hence the name "Z-form" (Figure 5). The new structure differed from the previously known forms of the double helix in a number of radical ways. First of all, as already mentioned, the Z-helix is left-handed, whereas both the A-form and the B-form are right-handed. Secondly, the Z-form's repetitive unit is a twosome of adjacent base pairs rather than a single base pair. The *syn* and *anti* conformations of the nucleotide links alternate in each complementary strand, while in each base pair one base is always *syn* and the other *anti* relative to the glycoside bond. This is diagrammatically shown in Figure 6. The winding angle (i.e., a pair's rotation relative to its neighbor around the helix axis) is –9° or –51° for Z-DNA (depending on whether it is an *anti-syn* or a *syn-anti* contact), as compared with 34° for B-DNA. Thus, there are 12 bp to a Z-helix turn.

Z-DNA has some other distinctive features. Each complementary base pair has two sides. In Z-DNA, the base pairs are inverted relative to the direction of the sugar-phosphate chains as compared with the B-helix. Sugar conformations also alternate along each chain. A nucleotide's *anti* conformation corresponds to 2'-*endo* sugar, while a nucleotide's *syn* conformation corresponds to a 3'-*endo* sugar. Z-DNA's mean inter-pair distance along the helix axis is 0.37 nm. The reader can find a detailed description of Z-DNA in the review in Rich et al.[4]

Generally speaking, the Z-form can be realized for any nucleotide sequence. Some sequences, however, show a definite predilection for it over others. As already mentioned in Section I.A, the *syn* conformation of nucleotides is almost equivalent energy-wise to the *anti* conformation for purines, while it is less advantageous for pyrimidines. Therefore, the Z-form must clearly arise in sequences where purines and pyrimidines regularly alternate in each strand. In this case, the Z-helix can be (and actually is) phased sequence-wise in such a way that all the purines are *syn* and all the pyrimidines are *anti*. It was precisely for such a sequence that the left-handed double helix was discovered.

D. Other Regular Structures

Apart from double helices with Watson-Crick base pairs, other regular structures may occur under certain conditions for certain nucleotide sequences. At acid pH, double helices may form with protonated pairs of two poly(A) chains or two poly(C) chains (these structures only been observed for ribopolynucleotides so far).[1] We are currently more interested in triple helices, which arise in circular DNA under certain conditions. Thus, a triple helix made up of two poly(dT) chains and one poly(dA) chain has been observed experimentally.[1] Within this helix, poly(dA) and one of the poly(dT) chains form a regular double helix, whose structure comes close to the A-form, while the other poly(dT) chain is embedded in the double helix's nonglycoside groove. The second thymine forms additional hydrogen bonds with the AT pair. A similar structure was proposed for a three-stranded complex of two poly[d(CT)] chains and one poly[d(AG)].[5] In this case, base triplets are formed, TTA and CGC, with the latter being protonated. The supposed structure of these triplets is shown in Figure 7. Because the CGC triplet is protonated, the complex can be stabilized by reducing the pH of the solution. No precise X-ray structure of the complex is available to date.

A triple helix can also be formed by two polypurine strands and one polypyrimidine strand, with GGC and AAT triplets.[6,7] In this case, the additional polypurine strand fits into the nonglycoside groove of the double helix. The supposed structure of these triplets is shown in Figure 8. Towards the end of 1989, the $d(G)_n$ oligonucleotides were reported to form a four-stranded helix under certain conditions.[8] Now we know that groups of four adenines can be integrated in a quadruple helix as well.[9] The supposed hydrogen bond diagrams for the GGGG and AAAA foursomes can be found in Lee.[9]

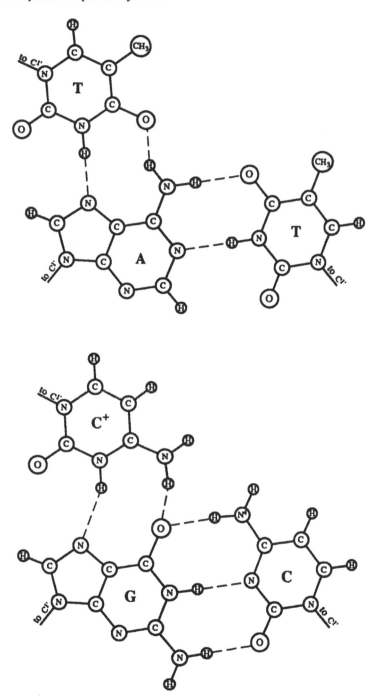

FIGURE 7. Proposed diagrams of hydrogen bonds in base triplets forming within poly[d(GA)] · ploy[d(CT)] · ploy[d(CT)] triple helices. In the CGC triplet, the cytosine forming a Hoogsteen pair with guanine is protonated.[5]

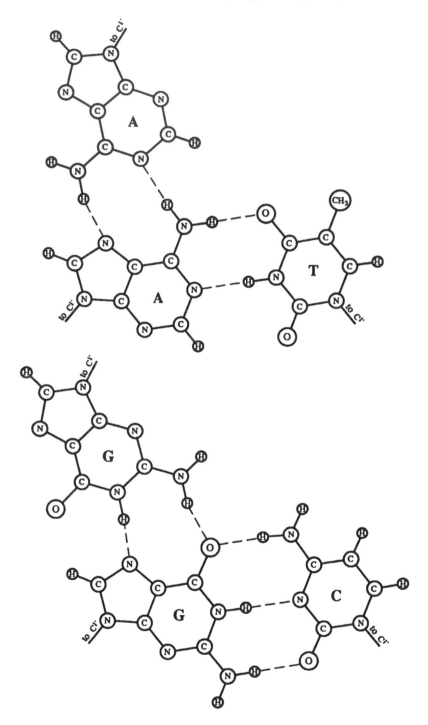

FIGURE 8.   Proposed diagrams of hydrogen bonds in base triplets forming within poly[d(GA)] · poly[d(GA)] · poly[d(CT)] triple helices.[6,7]

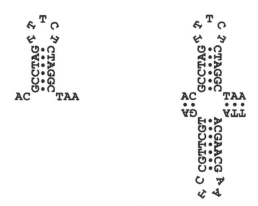

FIGURE 9. Hairpin and cruciform structures.

E. Hairpins and Cruciform Structures

Below a certain temperature threshold, a double helix made up by complementary strands is thermodynamically more advantageous than the loose state of those strands. Under these circumstances, single strands without a complementary partner also tend to form helical structures by stacking themselves into hairpins. Figure 9 shows what a hairpin looks like. Obviously, a hairpin cannot be formed by any nucleotide sequence. Equidistant nucleotides (in relation to the center) have to be mutually complementary. The loop at the hairpin's vertex must consist of no less than three nucleotides.

Hairpins are closely related to cruciform structures. A cruciform is presented in Figure 9. Such structures, in which hairpins consist of chains with different sequences, do not occur in native DNA molecules and can only be obtained artificially. They are stable, and the intersection point is rigidly defined by the sequence. Natural double-stranded DNAs frequently contain so-called palindromic sequences which have a second-order symmetry axis and are capable of forming symmetrical cruciform structures (Figure 10). In these cruciforms, the intersection point can shift with changing hairpin size. Palindromic regions can be imperfect, and then noncomplementary base pairs or bases pushed out of the helical areas must arise in the cruciform structure's hairpin. In linear DNA, the cruciform state of palindromic regions is less thermodynamically advantageous than the regular double-stranded form and can only occur in some cases as a metastable state. Stable cruciform structures of palindromic areas arise in circular DNA only.

II. FLEXIBILITY OF THE DOUBLE HELIX

A. Macromolecular Conformation in Solution

If one could look at the DNA double helix in solution, one would see that long enough molecules have the shape of loose coils instead of straight rods.

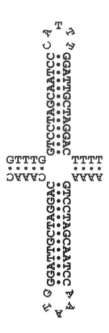

FIGURES 10. Cruciform and linear form of the ColE1 DNA palindrome. The symmetrical parts of the palindrome are highlighted in black type.

The reason for this is that the double helix structure is subject to thermal fluctuations. As a result of these fluctuations, adjacent links may differ in direction by several degrees, whereas their direction is strictly the same in an equilibrium structure. These random bends of the axis, which occur in every link of the chain, are poorly correlated. The average (over time) angle between the directions of two chain links is zero, even though there are exceptions to this rule for some special sequences (see Koo et al.[10] and the references contained therein). However, at any given time, this angle is, generally speaking, non-zero. As the distance between the links increases, so does the angle, so that a long enough DNA molecule assumes the shape of a coil.

The flexibility of the double helix is very important. Suffice it to say that if the DNA molecules had not been flexible, more often than not they would not fit into a cell. Flexibility is essential to the stacking of DNA in the chromatin and to the operation of many enzymes that interact with the double helix. To resolve a number of problems pertaining to the physical properties of DNA, one needs a quantitative measure of this flexibility and a physical model relating the flexibility of the double helix to its statistical, i.e., mean conformational, characteristics. These matters are treated by the statistical mechanics of polymeric chains. For a detailed survey of its foundations, the reader is referred to Cantor and Schimmel.[1] We shall confine ourselves to a brief summary of its basic propositions and then consider, in the next section, the statistical-mechanical model of the polymeric chain commonly used to describe the properties of the DNA double helix.

In the first approximation, the conformation of a polymeric molecule in solution can be described by the so-called free-jointed model. A free-jointed chain consists of straight segments of length l, each of which can equiprobably assume any spatial orientation regardless of the orientation of adjacent links. If such a chain consists of a large enough number of segments n, its characteristic conformation is a coil of size $\cong \sqrt{n} \, l$. This can be most easily demonstrated for the mean square distance between chain ends $<r^2>$, which is a good characteristic of chain size. By representing the chain's conformation as a set of vectors $\{l_i\}$, each corresponding to one segment, we obtain

$$\langle r^2 \rangle = \left\langle \left(\sum_{i=1}^{n} \mathbf{l}_i \right)^2 \right\rangle = \left\langle \left(\sum_{i,j=1}^{n} \mathbf{l}_i \mathbf{l}_j \right) \right\rangle = \left\langle \left(\sum_{i=1}^{n} \mathbf{l}_i^2 \right) \right\rangle + \left\langle \left(\sum_{i \neq j} \mathbf{l}_i \mathbf{l}_j \right) \right\rangle = nl^2 \quad (1.1)$$

and, consequently, $\sqrt{<r^2>} = \sqrt{n} \, l$. The expression in brackets $<>$ is averaged over all conformations. In deriving Equation (1.1), we used the fact that the scalar product of the vectors $\mathbf{l}_i \mathbf{l}_j$ averaged for all the directions of the vectors $\mathbf{l}_i$ and $\mathbf{l}_j$ is zero for $i \neq j$ for the simple reason of symmetry.

One can demonstrate that the distribution of distances between ends $W(r)$ for $n \geq 10$ is as follows (for example, see Cantor and Schimmel[1]):

$$W(r) = \left(\frac{3}{2nl^2\pi} \right)^{\frac{3}{2}} \exp\left(-\frac{3r^2}{2nl^2} \right) 4\pi r^2 dr \quad (1.2)$$

This distribution has a fairly sharp maximum which is attained for r values only slightly below the mean square distance between ends. The reason for this particular pattern is that the number of chain conformations with $r \cong \sqrt{n} \, l$ is many times larger than that of stretched-out conformations. For a detailed characterization of the free-jointed chain, see Volkenshtein.[11]

To assess the conformity between the free-jointed chain and a real polymeric molecule, one needs to establish the number of links in the polymeric chain, or its contour length, which corresponds to a segment of the free-jointed chain. This length is crucial to the conformational properties of the polymeric chain in solution; it is called the Kuhn statistical segment of a given polymer. The length of the statistical segment l_o depends on the nature of the polymeric chain and, in some cases, on the ambient conditions in solution. This length and the number of statistical segments in the chain n_o are determined by the following obvious relations (provided that $n_o \geq 10$):

$$l_o n_o = L$$

$$l_o^2 n_o = \langle r^2 \rangle \qquad (1.3)$$

Here L and $\langle r^2 \rangle$ denote the contour length and the mean square distance between the molecule's ends, which must be determined experimentally. Note that if the conformations of the real chain are indeed readily described within the framework of the free-jointed model, the size of the statistical segment, as determined by relations in Equation (1.3), will not depend on the length of the polymeric molecule. We shall discuss the size of the statistical segment for DNA in Section II.C.

Clearly, the free-jointed chain cannot be used to describe the properties of those DNA molecules that comprise just a few statistical segments. In a DNA molecule, bends are gradually accumulated, so the double helix cannot be regarded as a rectilinear stretch within one statistical segment. This is immaterial for many conformational characteristics of long molecules, but for short DNA molecules, one sometimes has to use a more realistic model that is more like the real double helix.

B. Chains with Correlated Direction of Adjacent Links

Consider a chain consisting of N identical links of length l in which the direction of the i-th link can deviate somewhat from that of the $(i + 1)$-th link. We will assume that the chain's rigidity, with respect to such local bends, has an axial symmetry, i.e., no matter which way the direction of the i-th link deviates from that of the $(i - 1)$-th link. The deviation has the same energy characteristics. Figure 11 shows a diagram of such a chain.

Let us find the mean square distance between the ends of this chain. By regrouping the terms of the double sum, taking into account the chain's homogeneity, we obtain

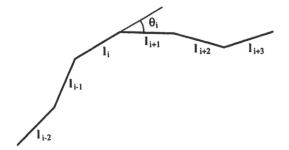

FIGURE 11. Chain with correlated directions of adjacent links. Conformation energy depends on the values of the θ_i angles.

$$\langle r^2 \rangle = \left\langle \left(\sum_{i=1}^{N} \mathbf{l}_i \right) \cdot \left(\sum_{j=1}^{N} \mathbf{l}_j \right) \right\rangle = Nl^2 + 2 \sum_{i=1}^{N-1} \sum_{k=1}^{N-i} \langle \mathbf{l}_i \mathbf{l}_{i+k} \rangle \qquad (1.4)$$

Let θ_i be the angle between the directions of the i-th and the $(i + 1)$-th links. Then,

$$\langle \mathbf{l}_i \mathbf{l}_{i+1} \rangle = l^2 \langle \cos \theta_i \rangle \qquad (1.5)$$

Since the chain is assumed to be homogeneous throughout, $<\cos \theta_i>$ does not depend on i. Let us introduce the symbol $<\cos \theta_i> = \alpha$. One can demonstrate (see Cantor and Schimmel,[1] for instance), taking into account the axial symmetry of deviations for adjacent links, that

$$\langle \mathbf{l}_i \mathbf{l}_{i+k} \rangle = \alpha^k \qquad (1.6)$$

The relation in Equation (1.6) allows the summation to be completed in Equation (1.4). Hence,

$$\langle r^2 \rangle = Nl^2 \left[\frac{1+\alpha}{1-\alpha} - \frac{2\alpha(1-\alpha^n)}{N(1-\alpha)^2} \right] \qquad (1.7)$$

Assuming N to be large enough, we can disregard the second term in brackets and, on the basis of Equations (1.3) and (1.7), obtain the following simple expression for the statistical segment of our chain.

$$l_o = l \cdot \frac{1+\alpha}{1-\alpha} \qquad (1.8)$$

If the average value of the angle θ_i tends to zero simultaneously with the segment length l, so that the value l_o is reserved, we arrive at the so-called wormlike chain. A fairly detailed description of this chain can be found in Cantor and Schimmel.[1] The rigidity of the wormlike chain is characterized by the parameter a, which is called persistence length and equals half of the Kuhn statistical length.

$$l_o = 2a \qquad (1.9)$$

Persistence length can be determined from the following relation:

$$\left\langle \left((\mathbf{l}_1 / l) \sum_{i=1}^{N} \mathbf{l}_i \right) \right\rangle = a, \qquad \text{for } N \Rightarrow \infty \qquad (1.10)$$

The left-hand part of Equation (1.10) corresponds to the mean projection of the vector linking the ends of the chain onto the direction of the first link. Thus, persistence length corresponds to the characteristic chain size for which the directions of elementary links are still correlated.

It turns out that the wormlike chain is an excellent approximation for describing DNA conformations in solution. The reason for this is that the DNA double helix is a very rigid polymeric molecule, and the mean angle between the direction of adjacent links (individual base pairs) is no more than a few degrees (see Section II.C).

The relation in Equation (1.8) allows the microscopic characteristic of rigidity $<\cos \theta>$ to be linked to the macroscopic one, i.e., the size of the statistical segment. Naturally, the value of $<\cos \theta>$ can be calculated for any microscopic model of bends involving adjacent links in DNA. Such models were first applied to DNA in by Schellman.[12]

The wormlike chain is very convenient for analytical calculations regarding the conformational properties of polymeric chains. At present, however, most theoretical studies of the conformational properties of polymeric chains are based on computer calculations in which the DNA double helix is simulated by a chain with finite mean angles between the directions of adjacent links. The simplest model of this kind assumes that the chain's energy E depends on the angles θ_i between adjacent links in the following way:

$$E = g \cdot \sum_i \theta_i^2 \qquad (1.11)$$

where g is the chain rigidity parameter. The number of links in such a chain that correspond to one statistical segment is determined by the relation in Equation (1.8). The value $<\cos \theta>$ in that relation depends on the constant g and is calculated according to the standard rules of statistical physics (for example, see Schellman[12]). Thus, having chosen a model in which k links correspond to persistence length, one can find the appropriate value of g on the basis of Equation (1.8). The conformational characteristic in question tends to some threshold value as k increases, i.e., as one approaches the persistence model. In real calculations, the minimum value of k required for an adequate description of the given characteristic is established experimentally. The computer approach to conformational analysis makes it possible to introduce various factors that bring the model closer to a real polymeric chain. One important factor that needs to be considered in a number of cases is the finite thickness of the chain. See Section II.D for a discussion of what this factor entails.

C. Persistence Length of DNA

Ways to determine the persistence length of the DNA double helix are discussed in a large number of papers. All these methods are based on measuring some characteristic that depends on the molecule's conformation in solution. The problem with finding persistence length experimentally has to do with the excluded volume effect which can affect the conformations of a polymeric chain in solution. What this effect is all about is the fact that no two chain links can occupy the same space point simultaneously. One result of the excluded volume effect, which is discussed in the next section, is that the mean square dimensions of a molecule depend both on its persistence length and on the excluded volume parameter. The first researchers who examined the value of a in the 1960s and in the early 1970s attempted different ways of allowing for the effect of excluded volume upon the coil size. A new approach has been evolved since the mid-1970s, wherein one measures the characteristics of short DNA molecules whose size does not exceed several persistence lengths. In this case, the likelihood of two chain links colliding in space is so low as to rule out the excluded volume effect. Almost all the recent studies have used DNA restriction fragments with a strictly defined length. Using short fragments of fixed length led to more precise and reliable results, so the matter of persistence length may be considered resolved.

For a relatively long time, almost all researchers have believed that, at a high ionic strength (above 0.1 M Na$^+$), the persistence length of DNA is about 50 nm (see the review by Hagerman[13]). There is the long-standing debate as to how the value of a changes with decreasing ionic strength. At present, it has been reliably demonstrated[14] that persistence length does not depend on ionic strength within a wide range of ionic conditions. Only at [Na$^+$] ≤ 0.002

M and at $[Mg^{++}] \leq 0.0003$ *M* do the polyelectrolyte properties of DNA begin to affect its flexibility. Since the negative charges carried by the double helix are not sufficiently screened, its bends require more free energy to be expended in this situation, so persistence length begins to increase. This ionic range is infrequently used in DNA experiments, so for all practical purposes, persistence length can be assumed to equal 50 nm.[13]

Using this persistence length value and the geometric parameters of a DNA link in the B-form, one can find, through Equation (1.8), that $\langle\cos\theta\rangle = 0.993$. Consequently, the characteristic values of the angle θ for the double helix are close to $\arccos(0.993) = 7°$.

D. Excluded Volume Effect

The previous model failed to take into account the fact that two chain links must always correspond to two physically different elements of space. Naturally, this must have an effect on the polymeric chain's conformation in solution, which is called the excluded volume effect or simply the volume effect.

In the case of the DNA double helix, the excluded volume effects are relatively small. This is due to the rigidity of the DNA molecule; its statistical segment length is 100 nm. Meanwhile, the geometric diameter of the double helix is close to 2 nm, i.e., the statistical segment is 50 times as long as it is thick. Thus, volume effects can often be disregarded for DNA, as indeed they have been in many studies. However, one can easily demonstrate that the volume effects in a polymeric chain must grow with increasing chain length (for example, see Tanford[15]). Therefore, volume effects should always be allowed for in the case of long enough chains.

What makes the situation somewhat more complex is the fact that DNA is a heavily charged polyelectrolyte. The negative charge carried by the double helix is almost completely screened at a certain distance from it by virtue of a particular pattern of low-molecular counterions in solution. The characteristic distance from the helix axis at which this screening is affected depends on the number and type of counterions in solution. If two chain segments are separated by a distance less than the characteristic screening radius, they should be subject to electrostatic repulsion, which increases the energy of the conformation in question, thus making it less advantageous. As a result, the effective excluded volume of a DNA segment is larger than its geometric volume. This, in turn, should augment the average size of the coil formed by the molecule.

It seems natural to characterize excluded volume by the effective diameter of the double helix, i.e., the diameter of a flexible impermeable cylindrical uncharged molecule that will have the same conformational characteristics in solution as real DNA. The matter of the effective diameter of DNA in solutions with different concentrations of Na^+ ions has been researched both experimentally[16,17] and theoretically.[18] The results of these studies show the effective diameter of DNA to increase from 4 to 6 nm at 0.2 *M* Na^+ to 15 to

22 nm at 0.01 M Na$^+$. However, the approaches used were not direct enough. The effective diameter of the double helix in solutions of varying ionic conditions needs further investigation.

E. Torsion Rigidity of the Double Helix

Apart from bending fluctuations, which exist in all polymeric molecules, the DNA double helix is characterized by torsion fluctuations. The latter only apply to such polymeric molecules where one part of the molecule cannot rotate around the chain axis relative to another part. As far as the DNA double helix is concerned, this property is ensured by the double-stranded structure. Single-stranded DNA does not have this property; it admits of easy rotation of chain parts relative to each other around single bonds in the sugar-phosphate skeleton.

The magnitude of torsion fluctuations is determined by torsion rigidity. The torsion rigidity constant corresponds to double the effort required to turn one end of a unit-length rod around its axis relative to its other end by 1 rad. If one knows the torsion rigidity constant, it is easy to find all the fluctuation characteristics of axial torsion between two given cross-sections of the molecule. The probability of the helical rotation angle deviating from the equilibrium one by the value $\Delta\alpha$ for the two cross-sections separated by distance L is

$$P(\Delta\alpha) = A \, \exp\left[-C(\Delta\alpha)^2 / (2Lk_BT)\right] \qquad (1.12)$$

where A denotes the normalization factor, k_B is the Boltzmann constant, and T is absolute temperature. Thus, the mean square amplitude of torsion fluctuations $\sqrt{<(\Delta\alpha)^2>}$ is defined by the equation

$$\sqrt{<(\Delta\alpha)^2>} = \sqrt{Lk_BT / C} \qquad (1.13)$$

If L is the distance between adjacent base pairs, we can use Equation (1.13) to find the fluctuation amplitude of the helical rotation angle in the double helix.

Two basic approaches are used for evaluating the torsion rigidity constant. One is based on studying the kinetics of fluorescence depolarization for a dye intercalated between base pairs in DNA. The kinetics of fluorescence depolarization is defined by the fastest movements of the DNA molecule in solution that can cause the dye molecules to change orientation. Analysis shows that torsion movements are just such movements.[19] A large number of studies based on this technique have yielded similar values for the torsion rigidity of DNA — around $1.5 \cdot 10^{-19}$ erg $\cdot$ cm (for example, see

Thomas et al.[20] and Millar et al.[21]). A value of C that comes close to this has been found through measurements of the correlation time of spin labels intercalated in DNA.[22]

A totally different approach to evaluating C is based on the properties of circular DNA. This approach will be discussed in detail in Chapter 3. Here we shall only note that the most reliable evaluations of C based on the properties of short circular DNA molecules come to $3 \cdot 10^{-19}$ erg $\cdot$ cm. This value of the torsion rigidity constant corresponds to the mean square fluctuation amplitude for a helical rotation angle of $4°$.

III. CONFORMATIONAL TRANSITIONS

A. General Properties of Conformational Transitions

Apart from minor fluctuations in the neighborhood of the principal conformational state under given conditions, DNA can undergo transitions between forms. These forms are separated in conformational space by energy barriers. Therefore, transitions between forms are different from smooth changes of equilibrium configurations within a form with changing ambient conditions. Three regular double-stranded forms of DNA are known at present: B, A, and Z. In addition, the double helix can be destroyed, leading to the disorderly form commonly referred to as the melted, or coil-like, state. In this case, the strands of the double helix form two single-stranded coils. Finally, certain structures can be formed by DNA segments with special sequences (see Section I.D and Section I.E.).

The equilibrium between different forms is shifted by changes in outside conditions, such as temperature, concentration of various ions, pH, solvent polarity, or the presence of various ligands in solution. Besides, the supercoiling of circular DNA is a special factor that can tip the balance between forms, but we are not going to discuss it here because this book has a whole chapter devoted to the effects of supercoiling on conformational transitions.

Here are the most important characteristics of conformational transitions in DNA. First, the equilibrium between forms depends on the nucleotide sequence. For instance, in the helix-coil transition, which occurs with a rise in temperature, the coil state is most readily realized for DNA segments enriched in AT pairs; the B-Z transition must happen first in the $d(GC)_n$ sequences, and so on. DNA is said to be a heterogeneous system with respect to conformational transitions. Second, the transitions are cooperative, i.e., in the transition range the DNA molecule breaks down into more or less lengthy regions in such a way that all the base pairs within a region are committed to the same form. The cooperativity of the transitions has to do with the fact that the emergence of boundaries between forms increases the free energy of the entire molecule.

The greater the DNA molecule's heterogeneity with respect to a particular conformational transition, the shorter (on an average) the regions into which it breaks down in the transition range. Cooperativity has the reverse effect. The larger the free energy of the boundaries between two DNA forms, the longer (on an average) these regions. The balance between these two factors determines the real molecule's breakdown in the transition range. The question of which regions in a given molecule and with what probability are in one or the other form can only be resolved on the basis of a statistical-mechanical treatment of the particular transition.

B. Statistical-Mechanical Treatment of Conformational Transitions

The statistical-mechanical treatment of conformational transitions in DNA is based on the analysis of equilibrium between two forms of DNA. The model describing this equilibrium assumes that each base pair can only be in one or the other state, e.g., the B-form or the A-form. We should note at once that these base pair states in fact combine a multitude of microstates that correspond to various conformational and fluctuational characteristics of a base pair (this is especially true of the coil or the open state) and to its diverse interactions with solvent molecules. Essentially, we assign all the microstates to two groups, and we shall only be interested in which of the two groups a base pair falls into. Obviously, this approach cannot claim to provide too detailed a description of the system, but the result has an attractive simplicity to it. Naturally, the microstates are combined into such groups that are readily distinguished by physical methods, so that the predictions of theoretical analysis can be subjected to direct experimental tests. This approach to the description of conformational transitions in biopolymers was first used by Zimm and Bragg.[23]

Now, consider a linear DNA molecule consisting of N base pairs, where each pair can be in one of two states, 1 or 2, which correspond to two forms of DNA. Clearly, the entire molecule can be in one of 2^N states. Since the equilibrium between these states is determined only by the free energy difference between them, we shall take our reading of free energy off the free energy of a molecule in which all links are in state 1. An important parameter that determines the equilibrium in the chain is the change in the free energy of the i-th link ΔF_i upon its transition from state 1 to state 2. More strictly speaking, the value ΔF_i corresponds to a process wherein the transition of the i-th link does not change the number of boundaries between regions with different states (Figure 12). The equilibrium constant of this process, which is equal to $\exp(\Delta F_i/RT)$, is denoted by s_i or ρ_i, depending on the form chosen as reference for the free energy reading. In this book, we shall read the free energy of different forms off its value in the B-form. Accordingly, we shall use the parameter ρ_i. The values of ΔF_i depend on the type of the i-th pair, i.e., in

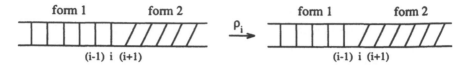

FIGURE 12.   Diagram of the 1 bp boundary shift between two DNA forms. The equilibriium constant of this process is equal to ρ_i.

the general case, they are different for the AT and GC pairs. In a number of cases, one should also take into account the dependence of ΔF_i on the type of adjacent pairs. Besides, the values of ΔF_i depend on the ambient conditions in solution (temperature, solvent composition, etc.), and it is this dependence that shifts the equilibrium between forms.

The other important parameter is the free energy F_j associated with the formation of a boundary between regions committed to different states. The formation of a boundary, or junction, between forms at any given point in the chain is always associated with an increase of free energy, i.e., the value of F_j is always positive. Normally, the value of F_j is assumed to remain unchanged in the range of transition between two DNA forms. Furthermore, the value of F_j is assumed to be independent of the types of base pairs forming the junction. These are legitimate assumptions, as the equilibrium between states 1 and 2 practically does not shift with small changes in the value of F_j. However, the very fact that the formation of a junction increases the chain's free energy has a very substantial effect on the transition, causing the cooperativity described in the previous section. The value $\exp(-2F_j/RT)$ is called the cooperativity factor and is usually denoted by σ. In this book, however, we shall use the symbol σ to denote superhelix density, and we shall introduce the value σ_{12} $= \exp(-F_j/RT)$ in lieu of the cooperativity factor.

The free energy of any state of the DNA molecule can be expressed through these two parameters, ΔF_i and F_j (within the framework of this simple model). For example, the free energy of a 4-bp segment with the sequence CGAC in state 2, surrounded by base pairs in state 1, is

$$\Delta F = 3\Delta F_{GC} + \Delta F_{AT} + 2F_j \tag{1.14}$$

By adding up expressions of the type in Equation (1.14) for all the link groups in state 2, we shall obtain the complete free energy of this state. The statistical-mechanical treatment of this system is attained in the standard manner by calculating the complete partition function Z over all the 2^N states:

$$Z = \sum_{k=1}^{2^N} \exp\left[-\sum_i \Delta F_i - 2n_k F_j\right] \tag{1.15}$$

The sum over i in this equation corresponds to summation over all the links in state 2; r_k denotes the number of segments consisting of links in state 2. The partition function can be expressed via parameters ρ_i and σ_{12}:

$$Z = \sum_{k=1}^{2^N} \prod_i \rho_i \hat{\sigma}_{12}^{2r_k} \qquad (1.16)$$

The product here is for all the links in state 2. The term of the partition function that corresponds to a specific state of the chain is called the statistical weight of that state.

Any statistical characteristics of the molecule can be expressed through the partition function. All the relations thus obtained are based on the fundamental fact that the probability of a system being in the i-th state equals the ratio of this state's statistical weight to the partition function. On this basis, one easily arrives at the conclusion that the probability of the chain's i-th link being in state 2, p_i, equals the ratio of the summarized statistical weights of the states in which the i-th link adopts state 2 to the complete partition function. This relation can be presented as follows:

$$p_i = \frac{\rho_i}{Z}\frac{\partial Z}{\partial \rho_i} = \frac{\partial \ln Z}{\partial \ln \rho_i} \qquad (1.17)$$

Consequently, the mean number of links in state 2, $<N_2>$, is defined by the equation

$$\langle N_2 \rangle = \sum_{i=1}^{N} p_i = \sum_{i-1}^{N} \frac{\partial \ln Z}{\partial \ln \rho_i} \qquad (1.18)$$

Thus, if we know the parameters ΔF_i and F_j and how ΔF_i changes with changing ambient conditions, we can use Equations (1.16) to (1.18) to obtain the complete description of the conformational transition in question. In the general case of an irregular DNA system, these calculations would require a computer. As a rule, the procedure does not involve going through all the states of the chain, but uses far more effective algorithms which are in mathematical terms equivalent to Equations (1.16) to (1.18). In a number of cases, however, the calculations can be made much simpler as long as the set of principal molecular states within the relevant range of ambient conditions can be designated beforehand. For example, if we have identified a short segment with a homogeneous sequence in which a given transition must happen much earlier than in the rest of the molecule, an approximate description of such a local transition can be based on a model allowing for just two possible states of the segment and of the molecule as a whole (either all the links of the segment are in state 1 or all the links of the segment are in state 2). In this case, all the statistical characteristics of the transition can easily be presented in a finite form (for instance, see Cantor and Schimmel[1]).

C. Helix-Coil Transition

The helix-coil transition is the best-studied conformational transition in DNA. It can be caused by various factors, but it is the temperature-related transition that gets the most attention. Therefore, it is also referred to as DNA melting. The transition can be monitored by a variety of techniques, as it entails changes in various physical properties of DNA (for instance, see Cantor and Schimmel[1]), but it is the changing absorption of light by a DNA solution in the wavelength range of $\lambda = 250$ to 270 nm. Upon the transition of DNA from helix to coil, the absorption of solution A in this wavelength range increases by 30 to 40%, and to find the mean degree of transition ϑ, i.e. the proportion of links that have adopted the coil state, one can use the equation

$$\vartheta = \left(A - A^h\right)/\left(A^c - A^h\right) \tag{1.19}$$

where A^h and A^c denote absorption in the fully helical and the fully coil states, respectively. Note that this method allows ϑ to be registered with an accuracy exceeding 0.1%. The melting of high-molecular DNA occurs within the temperature range of 3° to 20°, depending on the distribution of AT and GC pairs along the molecule. The simplest melting characteristic for this type of DNA is the melting temperature, T_m, which is defined as the temperature at which half the links in the molecule are in the coil state. For a given solvent, the melting temperature is in linear dependence on the proportion of GC pairs, x_{GC}, in the DNA:[1]

$$T = T_{AT} + \left(T_{GC} - T_{AT}\right)x_{GC} \tag{1.20}$$

where T_{AT} and T_{GC} denote the melting temperatures of DNA molecules made up entirely of AT pairs or of GC pairs, respectively.

By the mid-1970s, the DNA melting curve, i.e., the dependence of ϑ on T, was found to possess a fine structure if the length of the DNA molecule was not larger than tens of thousands of base pairs.[24] This fine structure is especially manifest on the differential melting curve, i.e., the dependence of $d\vartheta/dT$ on T. Figure 13 shows an example of such a differential melting curve for a fragment of fd phage DNA. The specific melting profile reflected by such curves is determined by the nucleotide sequence of the particular DNA under

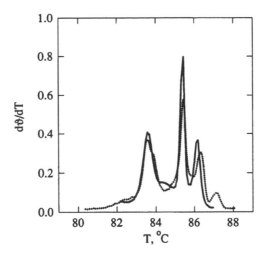

FIGURE 13. Experimental (····) and theoretical (—) differential melting curves for a 2528-bp-long fragment of phage fd DNA.

study. Theoretical[25] and experimental[26] analyses show that the peaks on the differential melting curves are associated with the melting of specific segments with a characteristic size of a few hundred base pairs within an interval of tenths of a degree. The discovery of the melting curve fine structure on the one hand and the decoding of long DNA sequences on the other hand allowed the theoretical description of the helix-coil transition to be critically tested. Though the statistical-mechanical model of the transition provided a good description of the fine structure effect itself, it was not clear to what extent theory was capable of predicting the melting profile for a particular DNA molecule. A direct comparison of theoretical and experimental melting profiles for DNA of up to several thousand base pairs showed that the statistical-mechanical model did in fact quite accurately describe the real melting of DNA. Figure 13 shows a fairly typical example of such a comparison. This success of the theoretical treatment has conclusively demonstrated that the statistical-mechanical description of conformational transitions in DNA, based on the propositions described in the previous section, closely corresponds to the real processes.

Experimental and theoretical studies of the helix-coil transition are discussed in a great many original papers and reviews[24,27-30] to which the reader is referred for detailed information.

D. B-A Transition

As the alcohol concentration increases in water-alcohol solutions, DNA passes from the B- to the A-form.[31] This transition does not involve any significant change in absorption; it is most readily registered by the changing

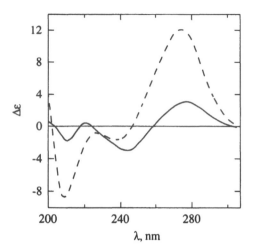

FIGURE 14.   Circular dichroism spectra for B-form (—) and A-form (---) DNA.

circular dichroism spectra in the near-ultraviolet (UV) region (Figure 14). The transition is only slightly dependent on the GC content of DNA and can be analyzed by means of the simplest statistical-mechanical model which assumes all base pairs to be equivalent with respect to this transition. The change in the free energy of a base pair upon its transition from the B- to the A-form in the vicinity of the transition range can be represented as[31]

$$\Delta F = (RT / Q)(a - a_o) \tag{1.21}$$

where a denotes the alcohol concentration in solution, a_o is the alcohol concentration at midtransition point, and $1/Q$ is the proportionality factor. According to Ivanov et al.,[32] $Q = 20$ and $a_o = 70\%$ for the trifluoroethanol solution (the most convenient for monitoring purposes). The parameter F_j for this transition is 1.8 kcal/mol. A typical B-A transition curve is shown in Figure 15. Within the framework of the model described above, the width of the transition, Δa, is related by a simple equation to the parameters Q and F_j:[31]

$$\Delta a = 4Q \cdot \exp(-F_j / RT) \tag{1.22}$$

There are no data at present as to the base pair free energy change attending the B-A transition in aqueous solutions. The formal substitution of zero value for a in Equation (1.21) yields a ΔF value of about 2 kcal/mol. However, such a far-reaching extrapolation of the results obtained for high alcohol concentrations is questionable.

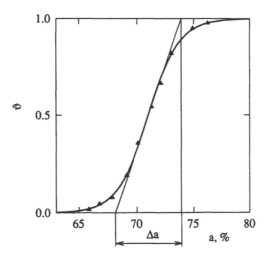

FIGURE 15. Dependence of the degree of B-A transition in DNA, ϑ, on the concentration of trifluoroethanol.[32] The transition width, Δa, is shown.

E. B-Z Transition

The Z-form was discovered for the hexanucleotide d(CGCGCG),[33] and indeed poly[d(GC)] is the type of polynucleotide that most readily undergoes the transition to the left-handed Z-form. Most researchers who have studied the B-Z transition in linear DNA used this polynucleotide or its methylated analogs (see Rich et al.[4] and the references contained therein). The transition of poly[d(GC)] from the B- to the Z-form occurs at a high concentration of sodium ions (2.4 M) or in the presence of certain bivalent ions. As with the B-A transition, the transition to the Z-form is best registered by the changing circular dichroism spectra in the near-UV region (Figure 16). Another method for registering Z-DNA is based on binding Z-form-specific antibodies (see Rich et al.[4]). A statistical-mechanical treatment of the B-Z transition in a polynucleotide can be accomplished by analogy with the treatment of the B-A transition in DNA, so we are not going to dwell on that here.

Circular DNA studies have demonstrated that the transition to the Z-form in arbitrary sequence DNA is not subject to any strict prohibitions. It is just that the emergence of this form of the double helix in the general case requires the consumption of large amounts of energy. For the Z-form to arise under given conditions, the energy has to be compensated in some way. This compensation can be attained through torsional stress (see Chapter 4) or through the specific binding of certain proteins. Now, to help the reader understand the formation of the Z-form, we shall consider the statistical-mechanical treatment of the B-Z transition in linear DNA with an arbitrary sequence.

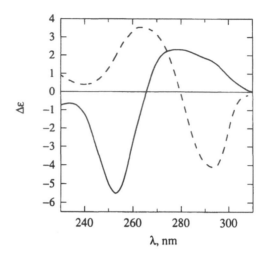

FIGURE 16. Circular dichroism spectra for B-form (—) and Z-form (---) poly[d(GC)] · poly{d(GC)].

In statistical-mechanical terms, the most important feature of Z-DNA is that the repetitive unit of its helix consists of two adjacent base pairs rather than a base pair. Therefore, any statistical-mechanical treatment should allow for the fact that each base pair in the Z-form can be in one of two energy-wise nonequivalent states.[4] The first (I) of these states is characterized by the *syn* conformation of purine and the *anti* conformation of pyrimidine. This state, which is more advantageous in energy terms, is realized in regular purine-pyrimidine sequences. The second state (II) corresponds to the *syn* conformation of pyrimidine and the *anti* conformation of purine. Though it is characterized by higher free energy, this state can be realized in some sequences because the regular Z-helix structure requires a strictly alternating pattern of *syn* and *anti* conformations in either nucleotide chain (see Figure 6). Meanwhile, in a regular urine-pyrimidine sequence, all the base pairs in the Z-form can be in state I, which is precisely why such sequences have been regarded all along as the most likely candidates for the formation of Z-DNA.[33] Thus, the free energy of a base pair's transition to the Z-form, ΔF, must depend not merely on the pair's type, but on its conformation (I or II) in the Z-helix. As usual, the additional free energy F_j^{BZ} corresponds to the boundary between two structural forms. However, another type of boundary can arise in the Z-form, the Z-Z boundary, which corresponds to a break in the regular *syn-anti* alternation pattern (since the sequence and conformation of nucleotides in one chain completely define the sequence and conformation in the other chain, we shall hereinafter consider one arbitrarily chosen chain only). This boundary, which can be regarded as a "phase change" in the Z-form, is associated with the free energy F_j^{ZZ}. By way of illustrating the free energy calculation for a segment in the Z-form, consider the example of a Z-helix formed by the GCGGTCC sequence and flanked by B-form segments.

Conformation of base in chosen chain	B	Z^s	Z^a	Z^s	Z^s	Z^a	Z^s	Z^a	B
Sequence of bases	A	G	C	G	G	T	C	C	A
Free energy of pair		ΔF_{GC}^{I}	ΔF_{GC}^{I}	ΔF_{GC}^{I}	ΔF_{GC}^{I}	ΔF_{AT}^{I}	ΔF_{GC}^{II}	ΔF_{GC}^{I}	
Free energy of boundary	F_j^{BZ}			F_j^{ZZ}				F_j^{BZ}	

Here Z^a and Z^s denote the *anti* and *syn* conformations of a nucleotide in the chosen DNA chain.

To sum up, this model allows for three possible states of each base pair and contains six energy parameters: ΔF_{GC}^{I}, ΔF_{AT}^{I}, ΔF_{GC}^{II}, ΔF_{AT}^{II}, F_j^{BZ}, and F_j^{ZZ}. Since the formation of Z-DNA is disadvantageous in energy terms under regular conditions in solution, all the values of ΔF must be positive (at high concentrations of sodium only ΔF_{GC}^{I} becomes negative). The parameters of the B-Z transition and the methods for evaluating them will be discussed at some length in Chapter 4. The above model of the B-Z transition was first proposed by Vologodskii.[34]

Chapter 2

CIRCULAR DNA AND SUPERCOILING

I. THE CIRCULAR FORM OF DNA

A. The Discovery of Circular DNA

In 1963, the DNA of the polyoma virus was found to exist in a closed circular form.[35,36] Two forms of circular molecules were extracted from the cell; they had different sedimentation rates and were designated as form I and form II. The more compact form I was found to turn into form II after a single-stranded break was introduced into one chain of the double helix. Subsequent studies performed by Vinograd et al.[37] linked the compactness of form I, in which both DNA strands are intact, to supercoiling. Form I came to be called the closed circular form. In this form, each of the two strands that make up the DNA molecule is closed in on itself. A diagrammatic view of closed circular DNA is presented in Figure 17. One can see the linkage of the two complementary strands.

For a while the discovery was not thought to be very significant, and closed circular DNA was dismissed as an exotic form. As time went by, however, more and more organisms were found to have it. Thus, for a number of phages (λ, φX174, and fd among others), DNA molecules that are present in phage particles in a different form were found to adopt the closed circular form during active functioning in the cell. At present, it is generally acknowledged that this form is typical of bacterial DNA and of cytoplasmic DNA in animals. Furthermore, giant DNA molecules in higher organisms form loop structures held together by protein fasteners in which each loop is largely analogous to closed circular DNA.[38-40]

Studies of closed circular DNA have shown many of this form's physical properties to be radically different from the properties of linear DNA. The difference is entirely due to the topological constraints that arise in the closed circular form. What these constraints come down to is that the topological state of the DNA strands must remain strictly unchanged through all the conformational transformations. Apparently, these topological constraints are removed as soon as a break appears in at least one of the strands. Therefore, closed circular DNA loses its special properties not only on being transformed into the linear form, but on turning into form II or an open (nicked) circular form in which one strand is broken while the other remains closed and circular.

FIGURE 17.   Diagram of closed circular DNA. The linking number of strands (Lk) is 17.

B. The Topological Aspect of Inter-Strand Linkage

The strands of the double helix are linked in the closed circular form. In topological terms, the links between strands belong to one class, for which there is a fairly simple and complete quantitative description (the inter-strand linkage will be discussed in a more general way in Chapter 3). The quantitative characteristic of such links is called the linking number and is determined in the following way. An imaginary surface is drawn up on one of the strands, then one calculates the algebraic (i.e., sign-dependent) number of intersections between this surface and the other strand. The result, denoted by two letters Lk, is called the linking number (the symbol α also stands for the linking number in some earlier papers on circular DNA). For closed circular DNA formed by a right-handed double helix, the linking number is considered to be positive. The linking number depends only on the topological state of the strands and is a topological invariant. Note the very important fact that the topological state (hence Lk) must be maintained through all conformational changes of closed circular DNA that occur without strand breaks. The highly distinctive conformational properties of closed circular DNA, as opposed to linear DNA, spring from here.

Quantitatively, the linking number is close to N/γ, where N is the number of base pairs in the molecule and γ is the number of base pairs per double helix turn in linear DNA under given conditions. However, this equation is not entirely satisfied, and the difference between Lk and N/γ plays an important part in the properties of closed circular DNA.

C. Linking Number Difference and Superhelical Density: Topoisomers

The fact that closed circular DNA has a topological invariant gives rise to a new physical parameter which characterizes this form and determines many of its properties. This parameter is called the linking number difference in closed circular DNA and is defined as

$$\Delta Lk = Lk - N / \gamma \tag{2.1}$$

Note that in many works published in the 1970s and 1980s, the parameter ΔLk was referred to as the number of supercoils and denoted by τ. There are two inferences to be made from the above definition.

1. The value of ΔLk is not a topological invariant. It depends on the solution conditions which determine γ. Even though γ itself changes very slightly with changing ambient conditions, these changes may substantially alter ΔLk, as the right-hand part of Equation (2.1) is the difference between two large quantities that are close in value.
2. The linking number is by definition an integer, whereas N/γ should not be an integer. Hence, ΔLk is not an integer either. However, the set of ΔLk values for a closed circular DNA with a prescribed chemical structure is always an arithmetical progression with a unit difference. This simply follows from the fact that, whatever the prescribed conditions, all changes in ΔLk can only be due to changes in Lk, since the value of N/γ is the same for all molecules. And of course any change of Lk would involve a temporary violation of the integrity of a double helix strand. Molecules that have the same chemical structure and differ only with respect to the linking number difference are called topoisomers.

These traits of the parameter ΔLk are explicitly or implicitly used in almost every experiment involving closed circular DNA. A clear understanding of these properties is therefore essential.

It often proves more convenient to use the value of superhelical density σ. Superhelical density is the linking number difference normalized for the length of circular DNA. This parameter is defined by the formula

$$\sigma = \gamma \cdot \Delta Lk / N \tag{2.2}$$

and equals the ratio of the linking number difference to the number of turns in a relaxed double helix.

Whenever $\Delta Lk \neq 0$, closed circular DNA is said to be supercoiled. Clearly, the entire double helix is stressed in the supercoiled condition. This stress can either lead to a change in the actual number of base pairs per helix turn in closed circular DNA or cause regular spatial deformation of the helix axis. The axis of the double helix then forms a helix of a higher order (Figure 18). It is this deformation of the helix axis in closed circular DNA that gave rise

FIGURE 18.   Diagram of supercoiled DNA. The double helix axis forms a helix of a higher order.

to the term "superhelicity" or "supercoiling". Native DNA extracted from cells turns out to be supercoiled, as well as closed circular. Supercoiling is always (or nearly always) negative ($\Delta Lk < 0$).

II. EXPERIMENTAL STUDIES OF SUPERCOILING

A. Determining the Linking Number Difference

At present, there are two fundamentally different methods for determining the linking number difference in a closed circular DNA preparation. The first method is based on the titration of supercoils by an intercalating dye (see Bauer and Vinograd[41]). This approach is applicable only to negatively supercoiled closed circular DNA.

The molecules of some dyes have the ability to intercalate themselves between base pairs in the double helix on binding to DNA, thus reducing the helical rotation angle between the base pairs. As an increasing number of ligand molecules bind themselves to negatively supercoiled DNA, the tensions within the DNA molecule are gradually removed and the double helix axis "straightens out". This causes a decrease in the molecule's mobility, which is easily registered experimentally. Then, after the number of bound ligand molecules per base pair, v, exceeds the value

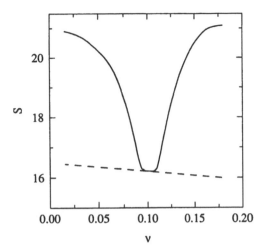

FIGURE 19. Dependence of the sedimentation constant for closed circular DNA of the SV40 virus (—) and for the same DNA's linear form (---) on the number of ligand (ethidium bromide) molecules bound to the double helix per base pair.[42]

$$v = 360 \cdot \Delta Lk \, / \, \phi N \tag{2.3}$$

where ϕ corresponds to the change of angle between adjacent pairs upon the integration of a ligand molecule between them, the tensions in DNA will start growing again. This will increase the molecule's mobility. As a result, the curve for the dependence, say, of the sedimentation constant on the number of bound ligand molecules has a minimum (Figure 19). Having found the value of v at the minimum and substituted it into Equation (2.3), one can determine ΔLk as long as one knows the value of ϕ. For ethidium bromide, which is most often used for titrating supercoiled DNA, $\phi = -26°$. The number of bound ligand molecules can be found with the help of spectral methods.[41] The shortcomings of this approach are its labor intensity and the need to know the ϕ value beforehand. For a long time, the uncertainty regarding the ϕ value was the main obstacle in determining the superhelical density of closed circular DNA. The correct value of ϕ for ethidium bromide was first obtained by Wang in 1974[43] and proved to be roughly twice the value hitherto cited in literature. Therefore, the values of σ cited in works published prior to 1975 should be multiplied by 2.2.

In 1975, Keller proposed a fundamentally different approach to determining the linking number difference in closed circular DNA[44] based on the fact that the electrophoretic mobility of DNA is highly sensitive to even small changes in its conformational characteristics. Under appropriate experimental conditions,

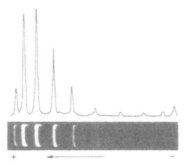

FIGURE 20.   Gel electrophoresis of pAO3 DNA. The arrow indicates the direction of mobility in the electrical field. Bottom: photograph of the gel. Top: densitogram of the same gel.

gel electrophoresis has such a high resolution that molecules with a Lk difference of 1 produce separate bands in the electrophoretic pattern. If a closed circular DNA preparation contains various topoisomers with the linking number difference ranging from zero to the maximum ΔLk value and they are all well resolved with respect to mobility, one can find the value of ΔLk corresponding to each band. One can then scan the gel to find the relative molecular content of each band and determine the mean value of ΔLk. The band that corresponds to $\Delta Lk \cong 0$ has the minimum mobility and can be identified through a comparison with the band for the nicked circular form. One should bear in mind the fact that topoisomer mobility is determined by the absolute value of ΔLk only, so the presence of topoisomers with positive supercoils can make interpreting the electrophoregram more difficult. By way of illustration, Figure 20 presents an electrophoregram for pAO3 DNA.[45] This DNA contains only 1683 bp, and for such short molecules the method's resolution proves exceptionally high. A major obstacle lies in the saturation of molecular mobility growth with increasing $|\Delta Lk|$. The greater the length of closed circular DNA, the sooner this saturation occurs.

A very elegant and effective way of increasing the method's resolution for large absolute values of ΔLk was proposed by Lee et al.[46] Here is a diagrammatic description of this approach. A preparation of closed circular DNA containing a mixture of topoisomers is applied to one corner of a flat gel, and electrophoresis is performed along one side of the gel. In this situation, the bands corresponding to topoisomers with large ΔLk values merge into one spot (for a DNA molecule of 4000 bp, this applies to topoisomers with $|\Delta Lk| \geq 15$). After that, the gel becomes saturated with the intercalating ligand whose binding to the closed circular DNA causes a partial relaxation of supercoiling stress. Now the mobility of topoisomers will no longer be determined by ΔLk, but by the value

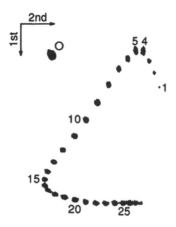

FIGURE 21. Separation of pBR322 DNA topoisomers by two-dimensional gel electrophoresis. Topoisomers 1–4 have positive superhelicity, the rest have negative superhelicity. After electrophoresis was performed in the first, vertical direction, the gel was saturated with ligand intercalating into the double helix. Upon electrophoresis in the second, horizontal direction, the 15th topoisomer turned out to be relaxed. The spot in the top left corner corresponds to the open circular form. (From Wang, J. C., Peck, L. J., and Becherer, K., *Cold Spring Harbor Symp. Quant. Biol.*, 47, 85, 1983. With permission.)

($\Delta Lk - N\nu\phi/360$), which can be regarded as the effective linking number difference. By choosing the appropriate ligand concentration in the gel, one can attain a good resolution of topoisomers that showed the same mobility prior to ligand binding. Now by performing electrophoresis in the perpendicular direction, we obtain the pattern shown in Figure 21. The number of topoisomers with $\Delta Lk \leq 0$ that can be resolved mobility-wise almost doubles in the case of two-dimensional electrophoresis. Besides, in this case, topoisomers with positive ΔLk have a different position in the electrophoretic pattern from topoisomers with the same negative ΔLk.

The separation of closed circular DNA topoisomers in a gel has proved to be one of the most potent techniques in DNA research. This method and its two-dimensional version have led to a whole series of remarkable experimental studies, some of which will be discussed below.

B. The Dependence of Superhelical Density on Solution Conditions

As mentioned above, the value of γ, which enters into the definition of linking number difference or superhelical density [see Equation (2.1)], depends on the ambient conditions in solution. Therefore, the linking number difference is not a constant value for DNA with a given Lk value. The effects of ambient conditions upon the linking number difference have been studied in some detail in experiments involving circular DNA. The reader will find a complete summary of those data in the review by Bauer.[47] We shall confine ourselves to quoting the key results of these studies.

The temperature coefficient of the change in superhelical density, $\Delta\sigma/\Delta T$, was found in the works of Wang,[48] Upholt et al.,[49] and Depew and Wang.[50] The results of these three studies are in relatively good agreement with one another and yield the average value of $\Delta\sigma/\Delta T$ as $3.1 \cdot 10^{-4}$ degrees^{-1}. The positive value of this coefficient means that σ increases, i.e., the double helix unwinds with rising temperature. This factor proves to be rather large in terms of its absolute value; a temperature change of $30°$ causes σ to change by approximately 0.01. The latter value constitutes about 20% of the characteristic superhelical density for a closed circular DNA isolated from the cell. Thus the temperature dependence of σ cannot be ignored in quantitative experiments involving circular DNA.

Another factor that has a considerable effect on the σ value is the changing ionic environment in solution. As a rule, an increase in ionic strength leads to a decrease of σ, i.e., the double helix winds itself up. In this situation the coefficient $\Delta\sigma/\Delta pX^+$ is positive. The most accurate data on the influence of various ions upon the σ value were obtained by Anderson and Bauer[51] (see also Bauer[47]). For univalent ions Na^+, K^+, Li^+, and NH_4^+ in the concentration range from 0.05 to 0.3 M, $\Delta\sigma/\Delta pX^+$ comes to $4.5 \cdot 10^{-3}$.

C. Superhelicity of Circular DNA Isolated from Cells

Closed circular DNA molecules isolated from cells almost always have a negative superhelicity. The values of σ for various plasmids and for phage, viral, and mitochondrial DNAs of animals usually lie in the range between -0.03 and -0.1. These data are summarized in the survey by Bauer.[47] The superhelical density of DNA isolated from cells largely varies ($\sigma = -0.03 \div -0.09$) even for plasmid and phage DNAs growing in the same host, *viz.*, *E. coli*.

Once the gel separation technique was developed, it became possible to examine the distribution of topoisomers in the preparation under study. This distribution, $P(\Delta Lk)$, proved to be rather wide.[52] It is fairly accurately described by the Gauss distribution function

$$P(\Delta Lk) = A \exp\left[-\left(\Delta Lk - \Delta Lk_o\right)^2 / 2r^2\right] \tag{2.4}$$

where A denotes the normalizing factor, ΔLk_o is the mean linking number difference in the preparation under study, and r^2 is the distribution variance. For a 5000-bp-long DNA molecule, the preparation contained about ten different topoisomers in comparable amounts. This is rather a lot considering that such DNA has only 20 to 30 supercoils on an average. Note that the topoisomeric distribution widths cited in Shure et al.[52] for DNA molecules isolated from phages, viruses, and plasmids proved to be 2 to 2.5 times larger than the widths of equilibrium thermal distributions for the same DNAs obtained through topoisomerase treatment (see Section III).

In recent years, however, it became clear that both the average value of ΔLk_0 and the distribution width, with respect to ΔLk depend on the life conditions of the cells from which the DNA has been isolated. For example, it was demonstrated that the topoisomeric distribution width of plasmid DNA isolated from logarithmically growing *E. coli* cells was considerably smaller than for DNA obtained from the same cells in the stationary phase.[53] The average superhelical density of plasmid DNA isolated from cells depends on the temperature at which those cells were grown.[53,54] Generally speaking, the superhelical density of a DNA preparation isolated from cells needs to be determined in each individual case.

D. Obtaining DNA with a Preset Superhelical Density

One often needs a DNA with a preset superhelical density. To obtain such a preparation, the following procedure should be used. Closed circular DNA molecules are treated with an enzyme called topoisomerase I. The action of this enzyme can alter the value of *Lk* in closed circular DNA as it introduces a single-stranded break into the double helix and then after a while "heals" it. In between these two events, the DNA may undergo changes in axial twist and shape (see Chapter 3), so that *Lk* gradually relaxes to its equilibrium value for the given conditions, i.e., the value corresponding to the minimum stress in closed circular DNA. This equilibrium value of can be changed within a wide range by adding various ligands which alter the helical rotation angle of the double helix upon binding to it. Thus, by adding varying amounts of an intercalating dye (usually ethidium bromide), one can obtain, after administering enzyme treatment and removing both ligand and enzyme from the solution, closed circular DNA with a certain required negative superhelicity. It is more difficult to obtain positively supercoiled DNA in this way, for no ligands are known to drastically increase the helical rotation angle on binding to the double helix. A small positive superhelicity was obtained by Snounou and Malcolm[55] with the help of netropsin.

Another possible method of altering superhelicity involves other enzymes, DNA gyrase or reverse DNA gyrase,[56] which create negative or positive superhelicity in closed circular DNA (see Section IV).

Naturally, neither of these two methods makes it possible to obtain closed circular DNA preparations with a strictly preset linking number difference.

Instead, they produce a certain distribution of ΔLk values around a mean value. The only way to obtain a preparation that comprises one topoisomer only is to separate a mix of topoisomers by electrophoresis and then to extract specific molecules from the appropriate section of the gel (for example, see Brown and Cozzarelli[57]).

III. SUPERCOILING ENERGY

Closed circular DNA molecules with a ΔLk value other than zero have additional free energy which is called supercoiling free energy. Supercoiling free energy must depend on the linking number difference and on the length of the DNA molecule. It can change with changing ambient conditions, such as the ionic environment and temperature. This section describes some experimental studies of supercoiling free energy. These studies provide the basis for the quantitative analysis of most problems considered in this book.

Supercoiling free energy was first determined in experiments involving the titration of closed circular DNA by an intercalating dye in Vinograd's laboratory.[58] The same approach was later used by Hsieh and Wang.[59] Supercoiling free energy was determined by Bauer and Vinograd[58] and Hsieh and Wang[59] based on the following considerations.

Consider the change in the free energy of a system consisting of a linear DNA molecule N base pairs length and a ligand solution with the number of bound ligands increasing by $Nd\nu$ (the following analysis is based on Hsieh and Wang[59]). This change can be presented as

$$dG = \left[\Delta\mu\ (\nu) - \mu_o - RT \cdot \ln(yc)\right]Nd\nu \qquad (2.5)$$

where $\Delta\mu(\nu)$ is the change in the chemical potential of the binding site upon ligand binding, $\mu_o + RT \cdot \ln(yc)$ corresponds to the chemical potential of free ligands (for instance, see Cantor and Schimmel[1]), y is the ligand activity, and c is the concentration of free ligands in solution.

Upon the binding of $Nd\nu$ ligand molecules to closed circular DNA, the change in free energy dG^* will be

$$dG^* = \left[\Delta\mu^*(\nu) - \mu_o - RT \ln(yc^*)\right]Nd\nu + (\partial G_s / \dot{c}\ )d\nu \qquad (2.6)$$

Here * means that the parameters in question refer to closed circular DNA. The last term in Equation (2.6) corresponds to the change in supercoiling free energy upon the binding of $Nd\nu$ ligand molecules. Assuming that the change

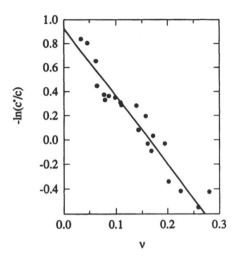

FIGURE 22.   Dependence of ln(c*/c) on ν for the binding of ethidium bromide to phage PM2 DNA. (From Hsieh, T.-S. and Wang, J. C., *Biochemistry*, 14, 527, 1975. With permission.).

in the chemical potential of the binding site upon ligand binding is the same in linear and closed circular DNAs as long as the degree of ligand binding, ν, is the same, i.e.,

$$\Delta\mu(v) = \Delta\mu^*(v) \tag{2.7}$$

the value of $(\partial G_s/\partial v)$ can be determined on the basis of Equations (2.5) and (2.6). Indeed, dG and dG^* are zero in the equilibrium situation, which leads us to equations corresponding to Equations (2.5) and (2.6). Now let the concentrations of free ligands in solutions be chosen in such a way that $v = v^*$. Then, by excluding the term $\Delta\mu(v) - \Delta\mu^*(v)$, we obtain

$$\partial G_s / \partial v = NRT \cdot \ln\!\left(c^* / c\right) \tag{2.8}$$

Equation (2.8) makes it possible to determine dG_s/dv from experiments involving the titration of linear and closed circular DNA by ligands, provided the values of ν and c can be found directly from the experiments [Equation (2.8) is valid only if $v = v^*$].

Figure 22 illustrates the experimentally found dependence of ln(c*/c) on ν. One can see that the experimental data are well described by the simple empirical equation

$$\ln\!\left(c^* / c\right) = a \cdot \left(v - v_c\right) \tag{2.9}$$

where the constant a, as demonstrated by Hsieh and Wang,[59] does not depend on the length of DNA. It follows from this equation that v_c corresponds to the linking number difference in the DNA preparation under study [see Equation (2.3)]. Integrating Equation (2.8), with Equation (2.9) taken into account, produces

$$G_s = NRTa / 2 \cdot (v - v_c)^2 \qquad (2.10)$$

In the absence of ligand molecules, i.e., for $v = 0$, if we take Equation (2.3) into account, we obtain

$$G_s = RTa(360)^2 (\Delta Lk)^2 / (2N\phi^2) \qquad (2.11)$$

Substituting the value 5 for a,[59] and $\phi = 26°$, we get

$$G_s = 500RT(\Delta Lk)^2 / N \qquad (2.12)$$

Note that an earlier study[58] had come up with a 1.5 times larger value for the coefficient preceding $(\Delta Lk)^2$ in the supercoiling energy formula.

A totally different approach to this problem was used in the very elegant experiments performed in 1975 at the same laboratories of Wang and Vinograd.[50,60] The authors of these studies examined the equilibrium thermal distribution of closed circular DNA over topoisomers. The distribution pattern was obtained with the help of topoisomerase I, which can alter the value of Lk in closed circular DNA by breaking and then restoring one strand in the double helix. Treating closed circular DNA with this enzyme causes the distribution of molecules with respect to Lk to relax towards the equilibrium form. The distribution of closed circular DNA molecules with respect to Lk was analyzed by means of the gel electrophoresis technique described in Section II.A. Figure 23 presents one of the electrophoretic patterns obtained in such experiments. Naturally, the maximum of the equilibrium distribution always corresponds to $\Delta Lk = 0$. It is also clear that molecules with close absolute values of ΔLk must be close in mobility, hence be next to each other in the gel. Distributions of the kind presented in Figure 23, where molecules having positive and negative ΔLk values are separated, are the result of electrophoresis conditions that differ from those used for the topoisomerase reaction. The change in conditions means that γ in Equation (2.2) needs to be replaced by a different value, γ', while Lk remains unchanged. As a result, the entire distribution will shift by the value $\Delta Lk - \Delta Lk' = N(1/\gamma - 1/\gamma')$, and for a large enough value of this difference all the toposisomers will be well separated.

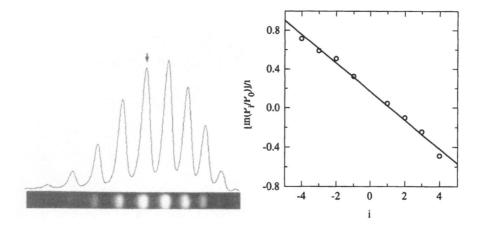

FIGURE 23. Equilibrium topoisomeric distribution of circular DNA PM2 (10,000 bp). We present a densitogram of the gel in which topoisomers were electrophoretically separated.[60] Adjacent peaks correspond to topoisomers that differ by unity in the ΔLk value.

The experiments performed by Depew and Wang[50] and Pulleyblank et al.[60] demonstrated that the resulting distribution is always a normal one. Most of the data published to date regarding the variance of this normal distribution for different DNAs,[50,52,60,61] $<(\Delta Lk)^2>$, are presented in Figure 24. The figure shows that the results obtained by different authors, who used quite different enzymic systems, fall within the same curve (this aggregate lot does not incorporate the results of Shore and Baldwin[62] which are somewhat different). This is a substantial argument in favor of the distributions being equilibrium.

The results of the experiments by Depew and Wang,[50] Shore et al.,[52] Pulleyblank et al.,[60] and Horowitz and Wang[61] satisfy the following phenomenological formula:

$$P(\Delta Lk) = A \exp\left(-K \cdot (\Delta Lk)^2 / N\right) \tag{2.13}$$

On the other hand, in view of the general relationship between the probability of a state and the corresponding free energy $G_s(\Delta Lk)$, an equilibrium distribution must have the form

$$P(\Delta Lk) = A \exp\left(-G_s(\Delta Lk) / RT\right) \tag{2.14}$$

It follows from the comparison of Equations (2.13) and (2.14) that

$$G_s(\Delta Lk) = KRT \cdot (\Delta Lk)^2 / N \tag{2.15}$$

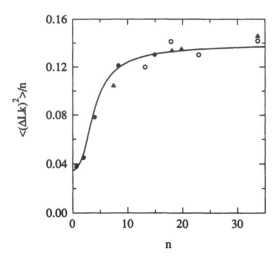

FIGURE 24.   Dependence of the equilibrium topoisomeric distribution variance, $<(\Delta Lk)^2>$, on the number of Kuhn statistical segments in DNA (300 bp equal to 1 Kuhn statistical segment). The variance values have been normalized for the number of segments. The experimental data are from Depew and Wang,[50] Shure et al.,[52] Pulleyblank et al.,[60] and Horowitz and Wang.[61]

or

$$G_s(\sigma) = KRTN\sigma^2 / \gamma^2 \qquad (2.16)$$

where K denotes the numerical coefficient defined by the formula

$$K = N / \left[\left\langle 2(\Delta Lk)^2 \right\rangle \right] \qquad (2.17)$$

As follows from Figure 24, for $N \geq 2500$ and $K = 1100$. Note that this value of K is considerably larger than that found in experiments involving an intercalating dye [see Equation (2.12)]. The difference is most probably due to the fact that the dye experiments had been performed in highly concentrated salt solutions.

In spite of its intrinsic elegance, the method of finding supercoiling energy, based on the equilibrium distribution of topoisomers, is not without its shortcomings, some of them considerable. First of all, this method allows the function $G_s(\sigma)$ to be determined only for $|\sigma| \leq 0.01$, whereas it is the $-0.03 + -0.08$ range that is of the greatest interest in physical and biological terms. Furthermore, these experiments can only be carried out for a relatively narrow range of ionic conditions which allow the enzymes in question to be active.

Recent theoretical calculations of the function $G_s(\sigma)$ have demonstrated[63] that the quadratic dependence on σ must be complied with up to $|\sigma| \cong 0.06$ only under certain ionic conditions, and the value of $<(\Delta Lk)^2>$, hence that of K, must depend on the ionic conditions (see Chapter 3). At present, however, the generally accepted practice is to use Equations (2.15) to (2.16) with $K = 1100$ in a wide range of σ values and for a variety of ionic conditions.

IV. TOPOISOMERASES

Topoisomerases are enzymes that can alter the topological state of circular DNA. Since a topological state cannot be altered without a break in the chain, the work of these enzymes comes down to creating a temporary single-stranded or double-stranded break, pulling another, whole, segment of the chain through the break and then "healing" the break. As a result of this enzymic activity, the chains keep their integrity, but their topological state may change. At present, it has been established that the cell needs topoisomerases to solve a number of topological problems arising during the replication and transcription of circular DNA. Besides, they maintain a certain level of supercoiling in the cell, at least in the case of the protozoa.

Topoisomerases fall into two types, I and II, according to their action mechanism. The first topoisomerase of type I was discovered by Wang in 1969.[64] By now, topoisomerases of type I have been found in virtually every class of living organisms. These enzymes operate by introducing a single-stranded break into the double helix. Topoisomerases of type I can cause the relaxation of negatively, and sometimes positively, supercoiled closed circular DNA, i.e., reduce superhelical stress by changing Lk. Their activity requires no co-factors. By creating a single-stranded break, the enzyme becomes covalently attached to one end of the chain, storing in the newly formed bond the energy necessary for the subsequent reunification.

The first topoisomerase of type II, named DNA gyrase, was discovered by Gellert et al. in 1976,[65] when the gyrase was found to create negative superhelicity in closed circular DNA in the presence of ATP accompanied by ATP hydrolysis. That was the definitive proof that the negative superhelicity of DNA isolated from cells is not linked to freeing it of the proteins that lend it a certain tertiary structure in the cell, but is indeed a necessary property for the normal functioning of DNA. The current notion of topoisomerase types only emerged in 1980, when the gyrase[57] and phage T4 topoisomerase[66] were clearly shown to break both strands of the double helix simultaneously and then to draw another double-stranded segment through the break. This reaction is diagrammatically presented in Figure 25. The fact that this is indeed what happens was proved by elegant experiments performed by Brown and Cozzarelli,[57] based on the understanding that during the reaction presented in Figure 25A the linking

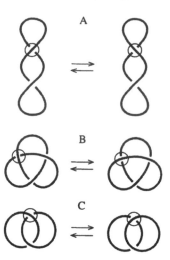

FIGURE 25.   Three "topological reactions" catalyzed by topoisomerases of type II. A — changing number of supercoils ($\Delta Lk = \pm 2$); B — unknotting and knotting of circular molecules; C — linking and unlinking of circular DNA molecules.

number, hence the linking number difference, must change by 2. Therefore, a topoisomerase of type II, acting on a certain topoisomer, could only turn it into topoisomers differing from the original one by an even linking number difference. That was precisely what the experiments[57,66] demonstrated. The fact that the linking number difference must change by 2 in the course of the reaction presented in Figure 25A follows from the curve's writhe properties (see Chapter 3). Nearly all reactions catalyzed by topoisomerases of type II require a concurrent hydrolysis of ATP.[67-70] The latter is essential to the supercoiling of closed circular DNA, which entails an increase of its free energy.

Until 1980, the changing superhelical density of closed circular DNA was believed to be the principal reaction catalyzed by topoisomerases. It suddenly transpired, however, that these enzymes can be responsible for other processes which alter the topology of the DNA chains. Topoisomerase of type II from the T4 phage was shown by Liu et al.[66] to catalyze the formation of knotted double-stranded circular DNA. The proof of the formation of knotted molecules by this enzyme was based on the electrophoretic separation of a topoisomerase-treated sample into fractions after introducing single-stranded breaks into the closed circular DNA form. The introduction of such breaks leads to a situation where molecules can differ from each other only by the type of the knot formed by the double helix axis, but not by Lk. Hence, electrophoresis separates them according to knot type only. The resulting fractions were subjected to electron microscopy (for a more detailed description of those studies see Chapter 3,

Section IV.C). Thus, it was possible to separate different types of knots as well as knotted molecules from unknotted ones. The equilibrium in the knotting reaction could be shifted either way depending on the experimental conditions.

Soon it became clear[71-74] that topoisomerases of type II could create the linkage of several double-stranded circular DNA molecules, in some cases forming whole nets like those observed in kinetoplasts.[75] The fact that this reaction occurred in a situation where the equilibrium had been shifted towards the unlinking of circular molecules furnished the first unambiguous proof to the effect that the giant nets of circular DNA molecules in kinetoplasts are linked in a way that is purely topological.[76]

Recently, knots and linkages have also been found to occur in double-stranded DNA through the action of topoisomerases of type I, provided the circular DNA carries single-stranded breaks.[77,78] In this case, the enzyme binds itself to DNA at the single-stranded break point and creates a break in the other strand, through which a segment of the double helix is then drawn. Under these circumstances, the formation of knots and linkages in double-stranded DNA by topoisomerases of type I is not at variance with their action mechanism described above.

An enzyme was recently discovered in thermophilic archebacteria which was given the name of reverse gyrase.[79-82] Coupled with ATP hydrolysis, this enzyme can create positive supercoiling in closed circular DNA molecules. As to its action mechanism, reverse gyrase is a topoisomerase of type I.[80-82] The DNA of archebacteria seems to be positively supercoiled. So far this has been demonstrated for the circular DNA of a virus-like particle isolated from archebacteria.[83]

The studies of topoisomerases and their biological function are rapidly gaining ground. A number of surveys have been published in recent years which provide ample information about these amazing enzymes, and supply the reader with a detailed bibliography on the subject.[67-60,84,85]

Chapter 3

GEOMETRY AND TOPOLOGY
OF CIRCULAR DNA

This chapter deals at some length with the topological properties of circular DNA. To describe these properties we shall have to introduce a number of mathematical notions. We shall assume that the reader has no preliminary knowledge of topology. The chapter begins with a differential geometric analysis of DNA supercoiling, which is essential to the study of circular DNA.

I. RIBBON THEORY

We have already mentioned the fact that supercoiling leads to the deformation of the axis of the circular closed DNA. The double helix axis itself can form a helix of a higher order, hence the term "supercoiling". On the other hand, it is clear that supercoiling can be realized, at least partially, through altering the mean winding angle of DNA. Thus, supercoiling can be structurally realized in two ways: by deforming the molecular axis and by altering the twist of the double helix. This can be ascertained by means of a simple experiment involving a rubber hose. Take a piece of hose and a short rod that can be pushed into the hose with some effort. The rod can be used to join together the ends of the hose and thus rule out their reciprocal rotation around the hose axis. If before joining the two ends one turns one of them several times around the axis, i.e., if one twists the hose, it will shape itself into a helical band once the ends are joined. If one draws longitudinal stripes on the hose prior to the experiment, it will be clear that reciprocal twisting of the ends also causes the hose's torsional deformation. Obviously, there has to be a relation between the mutual twist of the ends and the hose's torsional and bending deformation. This relation is established by the mathematical theory of ribbons. Even though this theory has been developed by mathematicians independently from the problems of circular DNA, it lends itself very well to the analysis of this particular object. Indeed, there is a wide range of outside conditions for which the double helix can be regarded as a smooth ribbon whose edges pass along the sugar-phosphate skeletons of the double helix. Then the linking number of the double helix strands would correspond to the linking number of the ribbon edges, each of which forms a closed line. On this basis, we are going to refer to the characteristics of a closed ribbon in the next section.

The principal mathematical result discussed below was obtained by White,[86] who developed the ideas of Calugareanu.[87] White's result was published in a mathematical journal and did not come to the attention of those engaged in

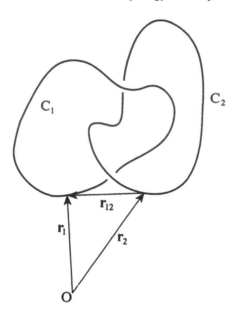

FIGURE 26. Notation of vectors in the calculation of the Gauss integral.

circular DNA studies. Therefore, it was so important that Fuller should write a paper 2 years later[88] and specifically suggest that this result be applied to the analysis of circular DNA. The same ideas were presented in a more popular way in an article by Crick which appeared in 1975.[89]

A. Writhe and Twist

Let us, therefore, assume that we have a closed ribbon whose edges form a linkage of a certain order. The problem is to find out how the linking number *Lk* is related to the differential geometrical characteristics of the ribbon, i.e., its twist and the spatial configuration of the ribbon axis. In Chapter 2, we defined the linking number as the algebraic number of intersections of the surface stretched on one contour by another contour. This definition is equivalent to the following mathematical equation:

$$Lk = \frac{1}{4\pi} \oint_{C_1} \oint_{C_2} \frac{(d\mathbf{r}_1 \times d\mathbf{r}_2)\mathbf{r}_{12}}{r_{12}^3} \tag{3.1}$$

where $\mathbf{r}_1$ and $\mathbf{r}_2$ are vectors whose ends run, upon integration, over the first and second contours, C_1 and C_2, respectively, $\mathbf{r}_{12} = \mathbf{r}_1 - \mathbf{r}_2$ (see Figure 26). The double contour integral in Equation (3.1) is called the Gauss integral. This integral is a topological invariant of the two contours; its value remains unchanged upon any mutual deformations of the contours that do not involve intersections of the strands, i.e., do not change their topological state. Its value can only be an integer. The proof of Equation (3.1) can be found in Edwards.[90]

The sign of the Gauss integral depends on the choice of directions along C_1 and C_2 and so is not defined for an arbitrary linkage. However, for the special case of so-called torus linkages (Section IV.A), the sign of the integral can be uniquely chosen to be the same assuming that the direction of both linked contours coincides. Then, for torus linkages corresponding to a right double helix, the linking number will be positive.

Though the properties of the Gauss integral are not too obvious, Equation (3.1) allows the mathematical apparatus of integral and differential calculus to be used for monitoring the Lk value, and this approach proves extremely effective.

It turns out that if the contours C_1 and C_2 are the edges of a smooth enough ribbon (i.e., such that the curvature radius at all points of the ribbon is far larger than its width), the integral in Equation (3.1) can be represented as the sum of two values designated as Tw and Wr. The first of these values is the ribbon's aggregate twist — a notion frequently used in differential geometry and in elasticity theory.[91] This value is expressed through a single-stage integral along the axis of the ribbon, C:

$$Tw = \frac{1}{2\pi} \int_0^L \left[\mathbf{r}'(s) \times \mathbf{a}(s) \right] \mathbf{a}'(s) ds \qquad (3.2)$$

This equation introduces the parametrization of the vector $\mathbf{r}$, which runs along the ribbon axis. The parameter s is the ribbon axis length reads off an arbitrarily chosen point; $\mathbf{r}'(s)$ denotes a derivative of the vector $\mathbf{r}(s)$ with respect to the parameter s. The vector $\mathbf{a}(s)$ characterizes the ribbon's local orientation relative to its axis; it lies within a plane tangential to the ribbon at the given point and perpendicular to the direction of its axis at that point. The vector $\mathbf{a}(s)$ is a unit vector. Finally, $\mathbf{a}'(s)$ is a derivative of the vector $\mathbf{a}(s)$, L is the complete length of the ribbon axis.

It is easy to see that, for a flat ribbon, the twist value is zero. Indeed, the vectors $\mathbf{r}'(s)$, $\mathbf{a}(s)$, and $\mathbf{a}'(s)$ lie within the same plane as the ribbon itself (see Figure 27). The vector $[\mathbf{r}'(s) \times \mathbf{a}(s)]$ is perpendicular to this plane, hence it is perpendicular to $\mathbf{a}'(s)$. Therefore, the scalar product of the vectors $[\mathbf{r}'(s) \times \mathbf{a}(s)]$ and $\mathbf{a}'(s)$ equals zero for all points on the ribbon axis. One can easily estimate the twist value for a straight ribbon twisted relative to the axis (naturally, the twist value has been defined for an open ribbon as well). In this case, it equals the total number of turns of the vector $\mathbf{a}(s)$ around the ribbon axis. In the general case, however, estimating a ribbon's twist can prove to be a difficult problem (see White and Bauer[92]). Still, twist, being an additive

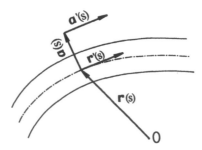

FIGURE 27.   Flat ribbon and the vectors defining its twist. Vectors $\mathbf{r}'(s)$, $\mathbf{a}(s)$, and $\mathbf{a}'(s)$ for such a band lie within a plane, and their mixed product $[(\mathbf{r}' \times \mathbf{a})\,\mathbf{a}']$ equals zero.

value, is a much simpler characteristic than linking number; the complete twist of a ribbon equals the sum-total of the twists of the ribbon's parts, as follows from Equation (3.2).

As it turns out, the linking number of a ribbon's edges does not, in the general case, equal its twist. The reason for this seemingly unexpected fact is that the value of Tw is the sum of minor turns of a vector perpendicular to the ribbon axis measured in a local system of coordinates, which itself turns with movement along the ribbon axis. The difference between the linking number of a ribbon's edges and its twist is equal to the writhe, Wr, value.

The Wr value is expressed through the Gauss integral, in which integration is performed both times along the same contour — the C ribbon axis:

$$Wr = \frac{1}{4\pi} \oint_C \oint_C \frac{(d\mathbf{r}_1 \times d\mathbf{r}_2)\mathbf{r}_{12}}{r_{12}^3} \tag{3.3}$$

The Wr value does not depend on the ribbon's orientation relative to its axis, but is defined entirely by the spatial course of the axis, i.e., is a characteristic of the closed curve. This new characteristic has been termed the curve's writhe.

Thus, the linking number Lk of the ribbon's edges can be represented as the sum of two values characterizing the ribbon's different degrees of freedom: twist around its axis, Tw, and deformation of the ribbon axis, Wr. This result can be presented in the following concise form:

$$Lk = Tw + Wr \tag{3.4}$$

This equation was first proved by White in 1969.[86] A less formal proof, which does not require the reader to have any special mathematical knowledge except familiarity with differential and integral calculus, has been proposed by V. V. Anshelevich. The reader will find it in the survey by Frank-Kamenetskii and Vologodskii.[93]

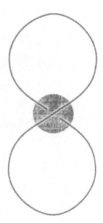

FIGURE 28. The near-flat "eight" figure. Only the dark-circled part of the contour juts out of the plane. The writhing of such a curve approaches −1.

B. Principal Properties of Writhe

Let us consider the most important properties of writhe. First of all, unlike the classical Gauss integral for two contours, which can only be an integer, a curve's writhe can have any value. It changes continuously with the curve's deformation that does not involve the intersection of segments. A curve's writhe does not change with a change in the curve's scale and depends solely on its shape.

As follows from Equation (3.3), the writhe value is zero for curves that have a center or a plane of symmetry. Thus, it is zero for any flat curve. It can be said that the writhe value is a measure of the curve's right-hand or left-hand asymmetry, i.e., a measure of its chirality.

Consider the writhe of an almost flat figure eight-shaped curve (Figure 28). Assume that in the area encircled by the broken line in Figure 28, contour segments come very close to each other (on the scale of the entire curve), but do not intersect. In this area, the curve slightly juts out of its "own" plane. As it turns out, the writhe of such a curve is −1 if the curve is a fragment of a right-handed helix (the case in Figure 28) and +1 if the crossing corresponds to a left-handed helix. This result does not depend on the shape or size of the two loops of the "8" as long as they do not project outside the plane or on the angle between the curve segments in the crossing area. This result follows directly from the definition of writhe in Equation (3.3), and we invite our mathematically inclined readers to prove it on their own. This result gives rise to another important property of writhe, which holds not only for the "quasi-flat" curve, but for an arbitrary curve as well. When the curve is deformed in such a way that one of its parts passes through another, the writhe value experiences a leap by ±2. This property has been used in analyzing the action mechanisms of topoisomerases (Chapter 2, Section V).

FIGURE 29.   Ribbon forming left-handed helix that clings tightly to the cylinder surface. The linking number of ribbon edges is −2. The writhing of the left-handed helix is negative.

By way of an example to illustrate Equation (3.4), consider a ribbon clinging tightly to the surface of a cylinder with its axis winding in the shape of a screw (Figure 29). Let R be the number of turns the ribbon axis makes around the cylinder axis, p the pitch of the helix, and r the cylinder radius. The sign of R is positive for a right-handed helix and negative for a left-handed one. It can easily be demonstrated that in this case

$$Lk = R \qquad (3.5)$$

The twist of such a ribbon can be calculated directly on the basis of Equation (3.2). The calculation shows[88] that

$$Tw = Rp \,/\, \sqrt{p^2 + 4\pi^2 r^2} \qquad (3.6)$$

The writhe of such a ribbon is most easily found on the basis of Equation (3.4). As follows from Equations (3.4) to (3.6), in this case,

$$Wr = R\!\left(1 - p \,/\, \sqrt{p^2 + 4\pi^2 r^2}\right) \qquad (3.7)$$

FIGURE 30.   The ribbon forms a double left-handed helix that clings tightly to the cylinder surface. The linking number of ribbon edges is zero. The writhing of the double left-handed helix is positive.

The same formula can be obtained directly by calculating the integral in Equation (3.3). Note that Equation (3.7) fits all the ribbons whose axes form a screw with pitch p, radius r, and number of turns R. This is not true of Equations (3.5) to (3.6), which are only valid for the special case of the ribbon clinging tightly to the cylinder surface. The same configuration of the ribbon axis, i.e., the same value of Wr, can correspond to a multiplicity of ribbons having different values of Tw and, consequently, Lk.

It follows from Equation (3.7) that the writhe of a screw line decreases with increasing p/r ratio, i.e., with the helix stretching out along the axis. By contrast, a double helical screw line (interwound helix) behaves in a very different manner if stretched out. Figure 30 shows a ribbon whose axis corresponds to such a line, wound around a cylinder. The writhe value of this line can be estimated on the basis of Equations (3.4) and (3.6). It is easy to see that for the ribbon clinging tightly to the cylinder surface the linking number is zero. The ribbon's twist can be determined with the help of the same formula in Equation (3.6), taking into account the fact that the ribbon in Figure 30 makes twice as many turns as the ribbon in Figure 29. Therefore, in this case, the writhe value is defined by the equation

$$Wr = -2Rp \,/\, \sqrt{p^2 + 4\pi^2 r^2} \qquad\qquad (3.8)$$

Note that in this case R corresponds to the number of helix turns in one half of the closed ribbon. It follows from Equation (3.8) that the absolute writhe value of the double helical screw line grows with increasing p/r ratio, i.e., the writhe of a double helix increases instead of decreasing upon stretching out, in contrast with a single helix. Furthermore, the left-handed (right-handed) single and double helices have opposite writhe signs (the writhe value of a right-handed double helix is negative).

II. CONFORMATIONS OF SUPERCOILED DNA

The chief result of the ribbon theory, expressed in Equation (3.4), is that supercoils in circular DNA can be structurally realized in two different ways. The first way consists in changing the twist of the double helix, the second lies in the deformation of the helix axis, giving rise to a certain writhe. Two questions naturally come to mind. First, what part of the linking number difference is realized through a changed twist and what part through writhe? Second, what conformations of the helix axis correspond to this or that writhe value in circular DNA? In this section, we are going to consider both the above and related questions.

General considerations suggest that the answers to these questions have to do with the relation between DNA's bending and torsional rigidities. While the bending rigidity of DNA has been studied for a long time and is known with reasonable accuracy (see Chapter 1, Section II.C), torsional rigidity was first determined through the analysis of circular DNA.[94] We shall dwell on this characteristic at some length in the next subsection. One significant point about this analysis is that it has been the first demonstration of how the ribbon theory can be applied to the quantitative analysis of the DNA double helix.

A. Torsional Rigidity of the Double Helix

The torsional rigidity of the double helix was determined on the basis of the equilibrium distribution of circular DNAs with respect to the linking number difference. This distribution was found experimentally for DNA molecules of different lengths and made it possible to determine the dependence of supercoiling energy on the linking number difference (see Chapter 2, Section III). Let us now consider those experiments in the light of the ribbon theory.

The experimentally observed[50,52,60,61] equilibrium fluctuations in the ΔLk value are made up of fluctuations in the twist of the double helix and in the writhe value Wr. In a circular DNA carrying a single-stranded break, the fluctuations of these two values occur independently from each other. The enzyme-assisted reunification of the DNA strand at the break point occurring at a random moment in time fixes the momentary value of the sum $Tw + Wr$, i.e., *Lk*. Therefore, the experimentally observed distribution of Δ*Lk* fluctuations in fact reflects the distribution of fluctuations in the values Δ*Tw* (Δ*Tw* = *Tw* − *N*/γ) and *Wr*. Clearly, the mean values <Δ*Tw*> and <*Wr* > are zero. However, the variances of these values, <(Δ*Tw*)²> and <(*Wr*)²>, which characterize their distribution widths in open circular DNA, are non-zero. Given the independence of the random values Δ*Tw* and *Wr* (we shall discuss this proposition later), the following equation must hold:

$$\left\langle (\Delta Lk)^2 \right\rangle = \left\langle (\Delta Tw)^2 \right\rangle + \left\langle (Wr)^2 \right\rangle \tag{3.9}$$

It was on the basis of this equation that torsional rigidity was determined.[94,95] Since the value <(Δ*Lk*)²> has been found experimentally (see Figure 24), and <(*Wr*)²> can be calculated (see below), Equation (3.9) makes it possible to determine <(Δ*Tw*)²> for any DNA. A simple equation relates <(Δ*Tw*)²> to the torsional rigidity of the double helix.

The problem of calculating the writhe variance for a closed chain was first formulated by Benham[96] and solved in Vologodskii et al.[94] Similar calculations were later reported by Le Bret[97] and Chen.[98] The method was based on a direct computer simulation of closed chain conformations and the calculation of the writhe value for each conformation. If the approved selection of closed chains correctly reflects the distribution of closed-molecule conformations in solution, this approach can be relied upon to give us a fair estimate of the statistical characteristic in question. The characteristic thus computed must obviously depend on the number of statistical segments in the simulated chain. Once we know the number of statistical segments in the circular DNA molecule under study, we can establish a correlation between the results of calculations for simulated chains and the writhe variance for the real DNA.

We are not going to dwell here on the choice of an adequate model of the polymeric chain or on the many interesting problems that arise during the calculation of this particular characteristic for closed chains (such as, for instance, the role of knotted closed chains). We shall only note that the excluded volume effects introduce an appreciable correction into the final result. The calculation of writhe variance and related problems are discussed in detail in Vologodskii et al., Frank-Kamenetskii et al., Klenin et al., and Shimada and Yamakawa[94,95,99-101] (see also Section V.B). The most complete data on the writhe variance for closed chains can be found in Klenin et al.[101] These data are presented in Figure 31. Allowing for the excluded volume effects does not change the pattern of the dependence, but only reduces the <(*Wr*)²> value for large *n* in accordance with the empirical relation

$$\left\langle (Wr)^2 \right\rangle = \left\langle (Wr)_o^2 \right\rangle / \left(1 + 6 \cdot d / l_o \right) \tag{3.10}$$

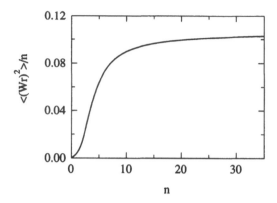

FIGURE 31.   Dependence of writhing variance on the number of Kuhn statistical segments n in a closed chain; $d/l_o = 0.02$. The curve is based on the data from Klenin et al.[101]

where $<(Wr)_o^2>$ denotes the writhe variance for an infinitely thin chain, d is the effective diameter of the double helix (see Chapter 1, Section II.D), and l_o is the length of the statistical segment of DNA.

As can be seen in Figure 31, the value $<(Wr)^2>/n$ grows fast at first with increasing chain length, then reaches a point of saturation. The experimental value $<(\Delta Lk)^2>/n$ shows a similar pattern (see Figure 24). One can see that specific traits appear in both figures in the same range of chain lengths. This is hardly surprising, as the values $<(\Delta Lk)^2>/n$ and $<(Wr)^2>/n$ must differ by the same constant for all chain lengths. This inference follows from an elementary analysis of the dependence of $<(\Delta Tw)^2>$ on n and from Equation (3.9).

Indeed, let the deviation of the winding angle between adjacent base pairs by $\Delta\varphi$ cause the free energy to increase by

$$f = \frac{1}{2}g(\Delta\varphi)^2 \tag{3.11}$$

where g is the torsional rigidity of the double helix per base pair (torsional rigidity C, introduced in Chapter 1, Section II.B, refers to a unit of helix length). Hence the distribution of $\Delta\varphi$ values must satisfy the equation

$$P(\Delta\varphi) = A\exp\left[-g(\Delta\varphi)^2 / (2RT)\right] \tag{3.12}$$

and the variance of this distribution is equal to RT/g. It is natural to suppose that the fluctuations of adjacent angles φ are independent, and, therefore, the

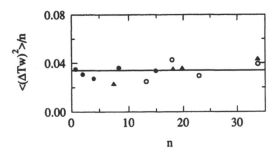

FIGURE 32. Dependence of twist variance on the number of Kuhn statistical segments in DNA. The points were obtained by subtracting from the experimental values of $\langle(\Delta Lk)^2\rangle$, cited in Figure 24, the calculated values of $\langle(Wr)^2\rangle$, cited in Figure 31.

variance of the twist for the entire molecule must be equal to the sum of the variances for individual angles of winding

$$\left\langle(\Delta Tw)^2\right\rangle = N\left\langle(\Delta\varphi)^2\right\rangle = NRT \,/\, g \qquad (3.13)$$

Thus, the twist variance must be proportional to the length of DNA. In full agreement with this conclusion, the difference between the experimentally determined value $\langle(\Delta Lk)^2\rangle/n$ (Figure 24) and the calculated value $\langle(Wr)^2\rangle/n$, equal to $\langle(\Delta Wr)^2\rangle/n$, as presented in Figure 32, proves to be independent from the length of DNA.

The twist variance found in this way made it possible to determine the torsional rigidity of the double helix on the basis of Equation (3.13). The numerical values of torsional rigidity are cited in Chapter 1, Section II.E.

According to the available data, the effective diameter of DNA, which is determined by how much the segments are electrostatically pushed apart, can be several times larger than the geometrical diameter of the double helix (see Chapter 1, Section II.D). It follows from Equation (3.10) that in those cases when the electrostatic diameter of DNA is more than twice as large as its geometrical diameter, the volume effects have to be taken into account in the calculation of writhe variance. Experimental data on the equilibrium distribution of ΔLk were obtained for such ionic conditions where the electrostatic effects must be relatively small (a high concentration of Na^+ ions or the presence of Mg^{2+} ions, which get very effectively bound to DNA), and the effective diameter of DNA should not be much larger than the geometrical diameter of the double helix. Hence, it is not so important to allow for the volume effects in the analysis of those experiments. However, quantitative analysis of the properties of circular DNA in solutions with a lower concentration of counterions absolutely requires that the volume effects be taken into account.

B. Computer Simulation of Supercoiling

The first attempts at a theoretical conformation analysis of supercoiled molecules were made within the framework of the elastic rod model.[102,103] When using this approach, one disregards entropy effects, i.e., ignores the issue of the number of close conformations realizing a certain writhe value. Attempts at an analytical treatment of this problem yielded results only for very small writhe values, where the conformation of a closed circular rod was not too different from a flat ring. Somewhat more significantly, Fuller carried out a pioneering qualitative analysis.[88] Fuller noted that for one possible conformation of the elastic rod which corresponds to a double helical line (Figure 30), the elastic energy per Wr unit, tends to zero as the ratio of the helix radius to the helix pitch decreases. Indeed, in this situation, the curvature of the line forming this superhelix tends to zero, while the writhe value tends to double the number of turns [see Equation (3.8)]. At the limit, the entire elastic energy will be concentrated in the two terminal loops of this structure. Of course, in a real molecule, the radius of the braid cannot be made smaller than the radius of the double helix, but the radius of the DNA double helix is small compared with the length of its statistical segment. Therefore, this is a likely conformation for supercoiled circular DNA. None of the other possible writhe realization forms has this property. For example, the reader can easily ascertain on the basis of Equation (3.7) that the elastic energy of a screw line wound on a torus (one of the alternative superhelix forms discussed in the literature) grows with increasing curve writhe.

A very successful computer approach to the problem is based on the search for the most probable conformations through the gradual deformation of a certain initial structure — a procedure widely known in statistical physics as the Metropolis method.[104] The success of this approach has been determined by the use of the very simple model of the DNA double helix described in Chapter 1, Section II.B. At first glance, the polymeric chain models discussed in Chapter 1, Section II do not seem to be suitable for the analysis of DNA supercoiling, as they simply fail to address the notion of twist. One can say that these chains have no torsional memory. However, this problem can be overcome with the help of the theorem expressed through Equation (3.4). While using the simple chain to simulate circular DNA, one can allow for torsional energy by expressing it as the difference between the linking number difference, ΔLk, and the writhe value, Wr, of the chain conformation in question. In this case, the linking number difference is an external parameter of the simulations. The energy of such a chain, with correlated directions of adjacent links (see Chapter 1, Section II.B), simulating circular DNA with the linking number difference ΔLk, can be represented as

$$E = g \cdot \sum_i \theta_i^2 + C / L \cdot (\Delta Lk - Wr)^2 \tag{3.14}$$

FIGURE 33. Typical conformations of supercoiled DNA obtained by computer simulation;[107] $\sigma = -0.03$.

where C/L corresponds to the torsional rigidity of a molecule of length L.

This approach was used by Hao and Olson[105] to determine the conformations of a supercoiled DNA with the minimum elastic energy (thus the problem of the conformation of supercoiled DNA was solved within the framework of the mechanical model; see also Tan and Harvey[106]). As mentioned above, statistical-mechanical analysis is much more adequate as an approach to a real situation in solution. Such an analysis of various properties of supercoiled DNA, based on the procedure described above, was carried out by Klenin et al.[63] and Vologodskii et al.[107] Let us consider the principal results.

Figures 33 to 35 present the typical simulated three-dimensional conformations of supercoiled DNA for different superhelix densities. One can see that, with growing superhelix density, the conformations become increasingly regular and can be described as an interwound superhelix with possible branchings. The question of the equilibrium number of branching points has been examined in detail by Vologodskii et al.[107] Calculations have also supplied an answer to the other question posed at the beginning of this section: the relative contributions of writhe and twist in the structural realization of supercoils. The ratio of the average writhe, $<Wr>$, to the linking number difference in DNA, ΔLk, depending on superhelix density, is presented in Figure 36. Clearly, about $3/4$ of the total linking number difference are realized in the form of writhe for molecules with more than 10 statistical segments (3000 bp). As the DNA length decreases, the share of writhe in the structural realization of supercoils drops dramatically for small $|\sigma|$ values and approaches zero for

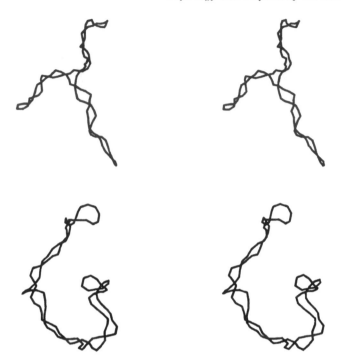

FIGURE 34.   Typical conformations of supercoiled DNA obtained by computer simulation;[107] $\sigma = -0.05$.

circular molecules whose length does not exceed one statistical segment. Furthermore, the results of the calculations depend on the effective diameter of the double helix. With decreasing ionic strength and the attendant increase of the effective diameter, the share of writhe decreases, and the mean superhelix diameter increases.[63,107] The data in Figure 36 correspond to a high-ionic-strength solution.

The theoretical approach developed by Klenin et al.[63] and Vologodskii et al.[107] provides answers to a wide range of questions pertaining to the conformational and thermodynamic properties of supercoiled DNA. Below we shall refer to some of the results obtained in these studies.

C. Experimental Studies of Supercoiled DNA Conformations

The most direct information regarding the conformations of supercoiled DNA is supplied by electron microscopy. The first such photographs were published in the pioneering paper by Weil and Vinograd back in 1963.[36] The very term "superhelix" was suggested by those first photographs that showed conformations corresponding to a twisted double braid. Later such conformations were observed by many authors, but it was only towards the end of the 1980s that a quantitative analysis was attempted. The equilibrium number of branching points in supercoiled DNA was examined by Laudon and Griffith.[108] That

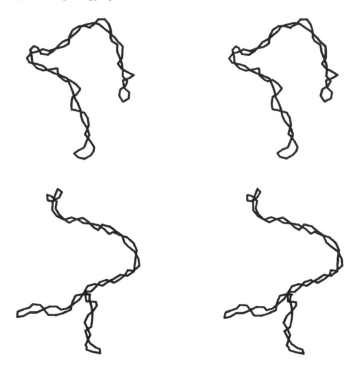

FIGURE 35. Typical conformations of supercoiled DNA obtained by computer simulation;[107] $\sigma = -0.07$.

study focussed on the effects that equilibrium bends of the double helix arising in certain nucleotide sequences have on the superhelix conformation. A detailed study of the superhelix and its characteristics was undertaken by Boles et al.[109] Figure 37 shows electron micrographs of supercoiled DNA obtained in that study. Generally speaking, the conformation of supercoiled DNA on a backing does not enable one to determine the writhe value for molecules in solution. The authors of this work circumvented this difficulty by using the data of electron microscopy and the analysis of site-specific recombination products. They assumed the length of the superhelix in solution to be the same as on the electron microscope backing. As a result, the authors came to the conclusion that the writhe of molecules came to about 70% of the ΔLk value, which is in complete agreement with the theoretical simulation results presented above (see Figure 36). Other data cited in Vologodskii et al.[107] also point to a good agreement between the results of electron microscopy and simulation. We should note, however, that the equilibrium numbers of superhelix branching points found by Laudon and Griffith[108] and and Boles et al.[109] were substantially different. Both experimental estimates in turn differed from the results of simulation.[107] This may indicate that the conformations of supercoiled DNA undergo substantial changes when a DNA preparation is applied to an electron microscope backing.

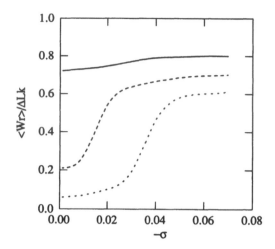

FIGURE 36.   Distribution of supercoiling between the change in twist, ΔTw, and the writhing value, Wr. We show the dependence of the ratio $<Wr>/\Delta Lk$ on σ for DNA molecules of more than 2500 bp (—), of 600 bp (---), and of 300 bp (····).

In this respect, cryoelectron microscopy has an indubitable advantage, as it allows molecules to be observed in vitrified water, i.e., in what amounts to a thin layer of solution. The results of a conformational study of supercoiled molecules by this method were recently published by Adrian et al.[110] Even though the micrographs (Figure 38) have a very low degree of contrast, they are projections of real three-dimensional conformations of DNA and so enable one to directly assess their writhe value. In complete agreement with the results cited above, Adrian et al.[110] found that about 70% of the ΔLk value is realized in the form of writhe. The same study revealed a very strong dependence of the superhelix diameter upon ionic conditions. While this diameter was about 12 nm at a low concentration of univalent ions, it dropped to 4 nm in a solution with 10 mM MgCl$_2$.

The conformations of supercoiled DNA have also been studied by other, less direct, experimental methods[111-113] based on hydrodynamic and spectrophotometric techniques. However, the results obtained by these methods were not as definite. These approaches can prove helpful in the studies of structural changes that may occur in supercoiled DNA under the influence of various factors.

D. Thermodynamics of Supercoiling

The free energy of circular DNA increases with supercoiling. Generally speaking, this can be caused by two things. First, elastic tensions, torsional and bending, which determine supercoiling enthalpy arise in supercoiled DNA. Second, superhelicity causes a decrease of the conformational freedom of the closed chain, i.e., a decrease of entropy. What is the relative contribution of

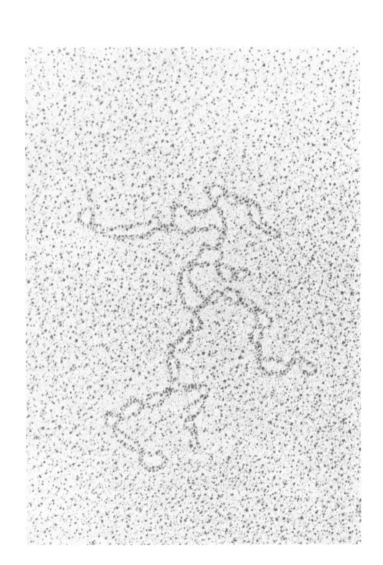

FIGURE 37. Electron micrograph of supercoiled DNA; the length of the molecule is 7 kb, $\sigma = -0.045$. Supplied by T. C. Boles.

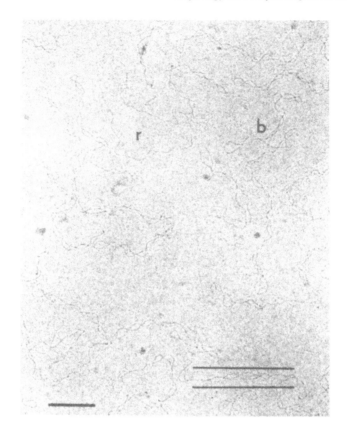

FIGURE 38.   Cryoelectron micrograph of naturally supercooiled pUC18 DNA (2686 bp).[110] The sample corresponds a low salt buffer. (From Adrian, M., ten Heggeler-Bordier, B., Wahli, W., Stasiak, A. Z., Stasiak, A., and Dubochet, J., *EMBO J.*, 9, 4551, 1990. With permission.)

these two factors to the free energy of supercoiling? In principle, this question can be answered by measuring the heat of supercoiling with the help of a microcalorimeter. Such an attempt was made by Seidl and Hinz,[114] who measured the emission of heat upon the relaxation of supercoiled DNA by topoisomerase I. Unfortunately, the results of that work are somewhat doubtful. According to those data, supercoiling entails an increase of not only the elastic energy, but also the entropy of circular DNA. Meanwhile, a simple comparison of the typical conformations of circular DNAs with different superhelical densities (see Figure 33 to 35) makes it all but certain that conformational entropy must decrease with the growing linking number difference.

The contribution of elastic tensions to the energy of supercoiling may be assessed on the basis of the computer simulation procedures briefly discussed in Section II.B. Figure 39 presents the results of such calculations obtained by Vologodskii et al.[107] The figure shows the dependences of the free energy

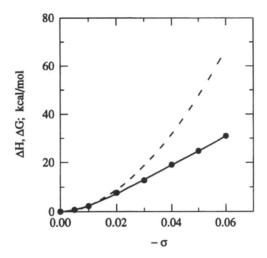

FIGURE 39. Dependence of the free energy of supercoiling (---), ΔG, and of supercoiling enthalphy (—), ΔH, on superhelix density $-\sigma$. Based on computer simulation.

and elastic energy of supercoiling upon superhelical density. Both relations were calculated for the same simulated chain, whose effective diameter corresponded to a high ionic strength. One can see that the contribution of elastic energy to the free energy of supercoiling decreases with increasing supercoil density: from 100% for very low σ down to approximately 50% for $\sigma = -0.06$. What this unexpected result means is that the number of available conformations in closed circular DNA is not restricted significantly by low supercoiling, but restriction becomes important as supercoiling increases. Meanwhile, there is clearly no reason why the quadratic dependence of the free energy of supercoiling, reliably established for a low superhelical density (see Chapter 2, Section III), should persist through high values of σ. Indeed, calculations[107] show that the deviations of the free energy dependence from the quadratic one prove to be considerable. Still, before any definitive conclusions can be drawn, one would need a detailed experimental study of the thermodynamics of supercoiling.

III. CYCLIZATION OF DNA

This section deals with the process whereby linear DNA molecules turn into circular ones. By now the cyclization of DNA has been well studied both theoretically and experimentally. Owing to these studies, a lot of our notions regarding the properties of DNA could be tested. In particular, direct and elegant evidence was produced to the effect that a double-stranded DNA in solution has a helical conformation with a helical repeat of about 10.5 bp per turn.

Suppose we have a solution of linear DNA molecules whose ends can be reunited and fixed in that state. The fixing may occur through enzymic covalent bonding between the 3′ and 5′ ends of the DNA chains. This reaction is catalyzed by DNA ligase. The reaction is considerably more effective if the DNA molecules have the so-called cohesive ends of the kind shown in Figure 40. A number of natural phage DNAs have cohesive ends. Shorter cohesive ends may arise after restriction endonucleases are applied to circular DNA molecules. When ends of this type stick together, complementary pairs are formed and are maintained in this bound condition for some time even without the formation of covalent bonds. The time it takes for them to fall apart largely depends on the number of base pairs formed and on the conditions in solution. If the length of cohesive ends exceeds ten nucleotides, the molecules can remain in the reunited condition for tens of hours and longer. The thermo-dynamics and kinetics of this process of circular DNA formation by molecules of 30 to 50 kb have been examined in detail by Wang and Davidson.[116-119] Based on these studies, let us consider the principal traits of this process.

First of all, let us analyze the equilibrium constant for the cyclization of DNA. The change in the molecule's free energy upon the transition from the linear form to the circular one, ΔF_o, is made up of the decreased conformational entropy of the entire chain upon cyclization (and the emergence of elastic tensions in the double helix for very short DNA molecules), and of the change in free energy of the cohesive ends when they form a helical structure, ΔF_{ch} (the coil-helix transition)

$$\Delta F_o = \Delta F_{ch} - RT \ln P_o \qquad (3.15)$$

where P_o is the probability of one chain end falling into the minor volume v in the vicinity of the other end. For a free-jointed chain of length L, the probability P_o is defined by the distribution function in Chapter 1, Equation (1.2) and is equal to

$$P_o = \left(\frac{3}{2\pi l_o L} \right)^{\frac{3}{2}} v \qquad (3.16)$$

where l_o is the length of the chain's statistical segment. Though the volume v cannot be determined independently, Equation (3.16) predicts quite a definite dependence of P_o on L, which is what this formula is about.

Equation (3.15) was tested experimentally Wang and Davidson.[116] The experiments were performed on phage λ DNA, which is diagrammatically

5' - GGGCGGCGACCTCGC • • •   48485 base pairs   • • • ACG
                     GCG • • •   • • • TGCCCCGCCGCTGGA - 5'

FIGURE 40. Cohesive ends of phage λ DNA.[115]

presented in Figure 40. That study demonstrated that the dependence of ΔF_0 on temperature and on ionic strength correlates very well with the dependence of ΔF_{ch} on the same parameters. Since the values of ΔF_{ch} are known from independent experiments for a wide range of ambient conditions (see Chapter 1, Section III.C), Equation (3.15) has made it possible to find the value of $RT \ln P_0$ by means of a comparison with the experimental data. Those calculations demonstrated that for 50°C and [Na$^+$] = 0.1 M,

$$-RT \cdot \ln P_0 = 14.4 \ \text{kcal} / \text{mol} \qquad (3.17)$$

The dependence of this factor on ambient conditions seems to be relatively slight. Therefore, Equations (3.15) to (3.17) make it possible to evaluate the cyclization constant for different DNAs under different ionic and temperature conditions.

The cyclization rate constant was determined for the same DNA for different temperatures and concentrations of sodium ions.[116] For [Na$^+$] = 0.13 M, the cyclization rate constant k enters into the following equation:

$$k = 3 \cdot 10^{13} \exp(-24,000 / RT) \ \text{min}^{-1} \qquad (3.18)$$

In the vicinity of the linear-circular transition range, the characteristic relaxation times determined by this constant run into tens of minutes. A comparison of the results for phage λ DNA[116] and for phage 186 DNA,[118] which has longer cohesive ends, shows that the cyclization rate strongly depends on the size of these ends.

It is very convenient to compare the cyclization and dimerization of the same DNA for the quantitative characterization of cyclization. Clearly, the cyclization of the original chains in solution must proceed concurrently with the formation of linear and circular dimers, trimers, etc. This process was first analyzed in the widely known study by Jacobson and Stockmayer.[120] The formation of linear dimers is characterized by certain rate and equilibrium constants for the bimolecular reaction in question. The ratio of the cyclization rate constant to the dimer formation rate constant, which is equal to the ratio of the equilibrium constants for the two processes, is called the *J*-factor.[121] For long enough chains, whose size exceeds several segments, the *J*-factor is simply equal to the concentration of one end of the molecule in the vicinity of the other end,

$$\left(\frac{3}{2\pi l_{o} L}\right)^{\frac{3}{2}}$$

Since both the rate constants and the equilibrium constants for cyclization and dimerization can be measured experimentally, one can also find the value of *J*-factor. In principle, this provides a method for determining the length of the statistical segment in DNA. One is better off using shorter DNA molecules than phage λ DNA for this purpose, so as to eliminate the excluded volume effects. However, for chains under 1000 bp, Equation (3.16) cannot correctly describe the likelihood of DNA ends approaching each other. One has to use the persistent model, instead of the free-jointed chain, to describe the conformational properties of such short molecules. The probability of the ends coming close to each other within this model was first calculated by Yamakawa and Stockmayer.[122] According to their results, the dependence of P_0 on L must reach a maximum when the chain length is roughly equal to one statistical segment; as L decreases further, the value of P_0 drops sharply. The reason for this is that the bending rigidity of DNA prevents the ends from coming together for such short molecules. The theoretical calculations[122] were compared with experimental data on the cyclization of short DNA molecules.[123] Figure 41 shows the results of that comparison. The agreement between experiment and theory, which carries no "fitted" parameters (the persistent length value was taken from an independent study), can be regarded as satisfactory. One thing that stands out, however, is the wide straggling of experimental points for the smallest DNA lengths. In 1983, Shore and Baldwin carried out a more thorough study of this length range and found the dependence of *J*-factor on DNA length to undergo oscillations with a period of about 10.5 bp (Figure 42). It is not a chance coincidence that this is also the helical repeat of the double helix. For such short molecules, the dependence of the torsional orientation distribution of the ends on the DNA length proves substantial. The torsional orientation distribution of the ends must have the same mean value and the same width as the equilibrium distribution of the same DNA with respect to topoisomers [see Equation (2.4)]. Unlike the topoisomeric distribution, however, the former is continuous, not discrete. Only certain narrow regions in this continuous distribution, with integer *Lk* values, correspond to a torsional orientation that admits of the chemical linking of the ends (see Figure 43). Two qualitatively different situations arise for long enough DNA and for very short DNA molecules. If the width of the torsional orientation distribution of the ends falls far short of 360°, the shift of the distribution center caused by a change in

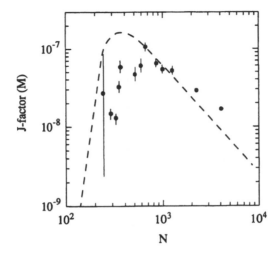

FIGURE 41. Cyclization probability for linear DNA molecules depending on their length. (From Shore, D. and Baldwin, R. L., *J. Mol. Biol.,* 170, 957, 1983. With permission.) We show the dependence of the *J*-factor measured in moles.

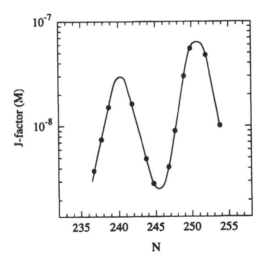

FIGURE 42. Cyclization probability (*J*-factor) for linear DNA molecules depending on their length, for chains of 237 to 254 bp. (From Shore, D. and Baldwin, R. L., *J. Mol. Biol.,* 170, 957, 1983. With permission.)

DNA length (the position of the center is determined by the mean aggregate winding angle of the double helix strands and shifts by $\cong 34°$ with a length change of 1 bp) must strongly affect the resultant probability of orientations that admit of cyclization. Naturally, the greatest probability of cyclization will

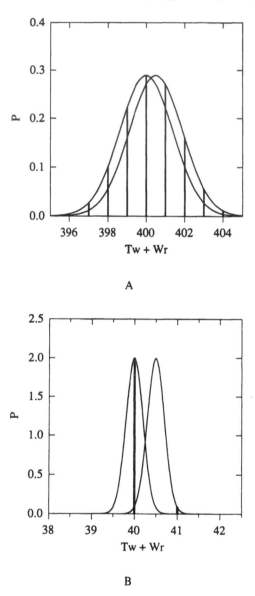

FIGURE 43.   Reciprocal torsional orientation of the double helix ends in the cyclization of a long (A) and a very short (B) DNA molecule. Continuous distribution of the ($Tw + Wr$) sum are shown by solid curves. At the moment of closing the ($Tw + Wr$) sum must be an integer; therefore, the closing probability is proportional to the sum of distribution values at integral points. The shift in distribution caused by an increase of the DNA length by 5 bp has no effect on the closing probability for long DNA, but changes the situation radically in the case of minirings.

be associated with those molecular lengths for which the distribution maximum corresponds to the torsional orientation of the ends favoring cyclization, i.e., a situation where the value of $34N/360$ approaches an integer (Figure 43B). By contrast, if the distribution width is far larger than 360°, the shift of the distribution center will not have any appreciable effect on the probability of orientations favoring cyclization (Figure 43A). Figure 24, which presents the dependence of $<(\Delta Lk)^2>$ on N, shows that the first situation corresponds to DNA molecules up to 500 bp long, while the second one applies to molecules of 1000 bp or more.

The data obtained by Shore and Baldwin[124] (see Figure 42) prove most elegantly and convincingly that DNA in solution has a helical structure with a helical repeat of 10.5 bp per turn. After Shore and Baldwin published their work, a number of authors attempted to use these data to find the persistent length of DNA and its torsional rigidity.[125,126] Those analyses of the experimental data were based on theoretical calculations of the probability of short double helices closing in upon themselves. This approach was most thoroughly realized by Hagerman,[127,128] who also reproduced Shore and Baldwin's experiments for short DNA fragments at various ionic strengths.[129] The resulting persistent length is in good agreement with the value found on the basis of other approaches (see Chapter 1, Section II.C). At the same time, torsional rigidity proves to be 1.5 times below the value yielded by an analysis of the topoisomeric distribution obtained in the same experiments (Section II.A). Though the exact causes of this discrepancy are not known, supposedly the analysis of cyclization kinetics requires a more detailed examination of the structure of circular DNA at the time of ligation, taking into account the possible stacking disorders at the junctions of the polynucleotide chains.

IV. KNOTTED AND LINKED DNA

As a linear DNA molecule links into a circle, it ends up in a specific topological state. Until now, we examined different topological states associated with the linking number governing the position of the strands of the double helix around each other. The strands' linking number, however, is not the only topological characteristic of circular DNA. As it links into a circle, any polymeric chain may finish up in an unknotted state or form a knot of some type or another. In the case of the DNA double helix, attention must be focussed on the type of the knot formed by the axis of the double helix (an unknotted closed chain is known as a trivial knot). With all conformational rearrangements of the chains, the type of the knot must remain unchanged. This factor must influence the conformational properties of closed chains, which must actually depend on the latter's topology. Knots often form in the course of various laboratory manipulations with DNA, though quite recently knotted DNA molecules were found also in living cells. It has turned out that analysis of the types of knots forming during the functioning of fermentation systems can help ascertain the mechanism of certain genetic processes involving DNA.

Just as different types of knots may form through the ring closure of single chains, a closed pair or a larger number of polymeric chains may form linkages of different types. One such linkage is formed by the strands of a double helix in a closed circular DNA form (in this chapter, we shall examine the DNA double helix mostly as an integral polymeric chain). Linked DNA molecules occur quite frequently in nature and can be obtained in lab conditions also. As it is, linkages of two chains, even as knots in isolated molecules, come in an infinitely large number of topologically nonequivalent types. The notion of a linking number, which was extensively used above, is good only for characterizing linkages of a certain class (torus linkages) forming in closed circular DNA. The overall picture is much more complicated, so before proceeding to the examination of concrete problems associated with knots and linkages in circular DNA, we shall undertake a general survey of the existing types of knots and linkages.

A. Types of Knots and Linkages

For the classification of knots and linkages, it must be agreed that any knot must be reduced to some standard type. One must be always careful, though, that there should be no self-intersection of chains in the course of such reduction. Deformations of this kind are known in topology as isotopic deformations. Two knots (or linkages) which can be transformed into each other by way of isotopic deformation belong to the same isotopic type. A standard type of knot (linkage) is such an image thereof when the minimum number of intersections on its plane projection is achieved. The simplest knot has three such intersections (Figure 44) and is called a trefoil.

Knots can be simple or composite. A knot is composite if there is an unlimited surface crossed by the knot at two points, which thus divides it into two knots. This definition is graphically presented in Figure 45. Figure 44 shows the initial part of the table of simplest knots — namely, all knots with less than nine intersections in the standard form. A knot and its mirror image are believed to belong to the same type of knot, although they may belong to the same or different isotopic types. In particular, the trefoil and its mirror image cannot be transformed into each other by way of continuous deformation without self-intersection, and, therefore, belong to different isotopic types. The figure eight knot (4_1) and its mirror image belong to the same isotopic type. The table of knots is, in fact, a table of knot types, for it features only one

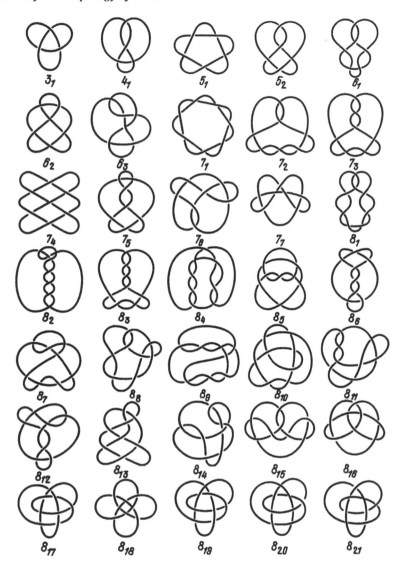

FIGURE 44. Table of simple knots with less than nine intersections in standard form.[130]

representative of mirror pairs. With the growth of the number of intersections, the number of types of simple knots grows very fast. Tables have been drawn up lately of all simple knots with less than 10 intersections,[130] with 10 intersections,[131,132] and so has a probably incomplete table of knots with 11 intersections (see Conway;[131] K. A. Perko, personal account). As it turns out, there are 49 types of knots with 9 intersections, 165 with 10 intersections, and

FIGURE 45. A composite knot. The shaded unbounded surface intersects this knot at two points only. By contracting these points into one along the surface, we obtain two knots of type 3_1.

about 552 types of knots with 11 intersections (pictures of all these knot types can be found in the studies cited above).

In contrast to the table of knots, the table of two-contour linkages was compiled very recently.[131] It contains 275 types of simple linkages with less than 11 intersections and is based on the same principle as the table of knots. The initial part of the table of linkage types is shown in Figure 46. It must be noted that of all the linkage types presented in the table, only linkages 2_1, 4_1, 6_1, and 8_1 belong to the torus class which corresponds to the linkages of the strands of the double helix in closed circular DNA (always granted that its axis is unknotted).

B. Theoretical Analysis of Knots and Linkages

The first question arising in the analysis of the topological properties of closed polymeric chains is what is the probability of different topological states occurring upon the spontaneous cyclization of chains in solution? In the case of isolated chains, this question is reduced to the probability of the formation of various knots at spontaneous locking. This question was clearly formulated by Delbruck[133] and resolved for the first time in the studies by Vologodskii et al.[134] and Frank-Kamenetskii et al.[135] The approach used in these studies is based on direct computer simulation of closed chain configurations and analysis of the topology of each such chain. Despite the seemingly obvious character of this approach, a whole number of serious problems emerge in the process, which are thoroughly examined in the review by Frank-Kamenetskii and Vologodskii.[93] We shall look into the most fundamental of these problems

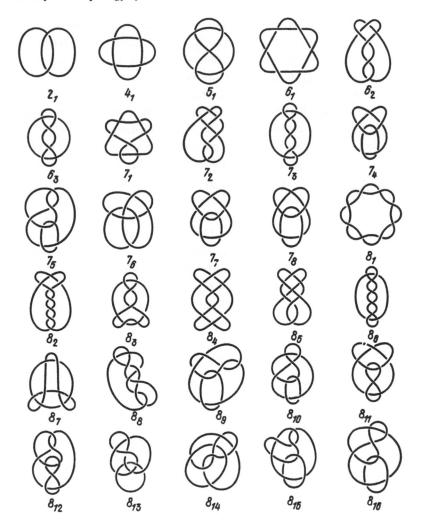

FIGURE 46. Table of links with less than nine intersections in standard form.[131]

in some detail here, for it is equally important for the analysis of the topological properties of all circular molecules.

The essential point about this problem is how to analyze the topology of a particular closed chain configuration, or how to determine the knot type to which this configuration belongs. Here is an example to illustrate this problem. Let us imagine that we have a heavily entangled circular length of rope rolled into a ball, and we want to find out whether it is knotted or not. The fact that persistent attempts to untangle the rope have produced no result cannot be taken as a proof that we are dealing with a nontrivial knot. What is needed for resolving this problem is a single-valued algorithm of verification of the topological identity of the configuration in question. The construction of such algorithms belongs to the realm of a branch of mathematics known as topology. In principle, all approaches to this problem are based on the construction of invariants of topological states. An invariant of the topological state is a characteristic thereof which remains unchanged with any deformations of the chain, which are possible without the disruption of their integrity. The simplest topological state invariant is the Gauss integral in Equation (3.1) which governs the linking number of two chains. Of course, for classifying the state of chains with a topological invariant, the latter must assume different values for different topological states. Not a single topological invariant meets this requirement in full measure, but there are very powerful ones among them which help identify many elementary types of knots (linkages) and distinguish them from more complex ones. The Gauss integral is a fairly weak topological invariant and is of no use for distinguishing many linked states of chains from unlinked states (the value of the Gauss integral is zero for unlinked chains and for 9 of the 30 simplest linkages shown in Figure 46). However, it identifies all linkages of the class corresponding to the linkages of double helix strands in closed circular DNA. In contrast to the Gauss integral, the Alexander polynomial (see reviews by Frank-Kamenetskii and Vologodskii[93] and Michels and Wiegel[136]) has proven to be a potent topological invariant, convenient for computer simulation. It is a polynomial of one variable in the case of knots and of two variables in the case of linkages. The calculation of the Alexander polynomial for any chain conformation can be done with a computer,[93,136] if the input data are duly adapted. It is precisely the invariant that was used in all studies dealing with the computer calculation of the topological properties of closed chains.[93,97,99,134-140]

Starting from 1974, about a dozen studies have been devoted to computer calculations of the probability of knotting arising upon random closure of the chain, P. Most simulations confined themselves to the analysis of an infinitely thin polymeric chain and studied the dependence of P on the number of statistical segments in the chain, n. All data obtained for an infinitely thin chain perfectly correlate with one another, while for chains containing more than 50 segments, they can be described well by the following empirical equation:[139]

$$P = 1 - \exp(-n / n_o) \text{ and } n_o = 266 \qquad (3.19)$$

It was noticed in the very first studies,[93,97] however, that the probability of the knot formation very sharply depends on the chain thickness. As revealed by detailed calculations,[99] the influence of the chain thickness is so great that it cannot be ignored even for such a molecule as the DNA double helix. The results of these calculations are presented in Figure 47. The figure shows that

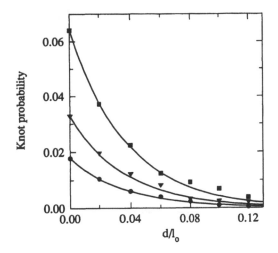

FIGURE 47. The probability of formation of a nontrivial knot upon random cyclization of a polymeric chain vs. the ratio of Kuhn segment's thickness d to its length l_o.[99] Data for a chain length consisting of 14 (●), 20 (▼), and 30 (■) Kuhn statistical segments are shown.

the probability of the formation of a cruciform drops almost tenfold upon the growth of the d/l_o ratio from 0.02 to 0.1. The conclusion is that experimental analysis of the dependence of the share of knotted DNA molecules occurring at their random cyclization in a solution may prove the most sensitive method of gauging the effective diameter of the double helix (see Chapter 1, Section II.D).

Recently, the question of the influence of the chain thickness on the probability of knotting was thoroughly studied for long chains[139] ($n = 30 \div 2000$). The findings of that study were likewise well described by Equation (3.19), but the n_o value increased extremely fast with the growth of the thickness of the model chain.

The studies[134,135] also provide data on the probability of the formation of knots of different types for the infinitely thin chain model.

On proceeding to linkages, we must first of all examine the question of the probability of the development of a linked (or nonlinked) state at spontaneous locking of two chains with a set distance, R, between the centers of their mass.[99,135,140,141] The results of such calculations for the infinitely thin chain model are presented in Figure 48. The considerable probability of the formation of linkages with small R values means that the number of states of a system of two nonlinked chains appreciably decreases as they come closer together. As a result, the solution of nonlinked infinitely thin closed polymeric chains will not be ideal. One of the reasons for this is inter-chain repulsion which

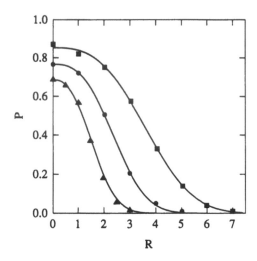

FIGURE 48.   The probability of a link formation between two chains vs. the distance between their centers of mass.[99] Different curves correspond to different numbers of Kuhn segments in each chain (both chains are assumed to have the same number of segments): 20 (▲), 40 (●), and 80 (■).

is of an entrophic nature. In statistical mechanics, such repulsion is quantitatively characterized by the second virial coefficient B.[142] The values of B for a solution of nonlinked rings can easily be calculated on the basis of the data presented in Figure 48. These values prove close to the B value conforming to mutually impermeable spherical particles which have a radius equivalent to the mean square radius of inertia of a closed polymeric chain.[99] In this manner, even ideal infinitely thin closed chains are bound to experience strong mutual repulsion which is wholly dependent on topological limitations.

In the event of chains of a finite width d, part of the mutual configurations of nonlinked chains get excluded for two reasons: because of topological limitations and because of the chains' intersections with one another.[99] The results of the calculations of the aggregate sum of forbidden configurations and of the share of forbidden mutual conformations caused by the intersection of chains are presented in Figure 49. The figure shows the dependence of the full second virial coefficient for such chains, B, and the share in this value of intersections, B_0, depending on the d/l_0 ratio. One can see that in this case, also, the impact of volume effects on topological characteristics is considerable even for very thin chains.

C. Knots and Linkages in DNA

The question of the possibility of the existence of knots and linkages in closed polymeric chains was first raised in the review by Frisch and Wasserman.[143] It was after the discovery of circular DNA, however, that the

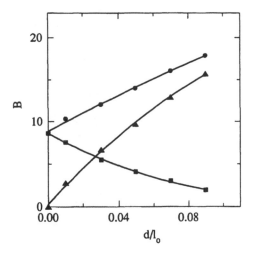

FIGURE 49. Dependence of the second virial coefficient B on the ratio of a Kuhn segment's thicknes d to its length l_o for closed chains of 20 segments (●).[99] Also shown are the contributions of volume (▲) and topological (■) effects to the value of B.

topology-related problems acquired special relevance. Soon after the discovery of solitary circular DNA molecules, catenanes[144,145] or linkages were detected in some cells. One absolutely amazing example of a topological structure is provided by the giant networks of linked circular DNA of kinetoplasts (see the review by Borst and Hoeijmakers[75]). These networks are comprised of tens of thousands of circular DNA molecules.

The principal methods for studying the topology of double-stranded DNA are electron microscopy and gel electrophoresis. On an ordinary electron microscopic photograph of DNA, however, it is pretty hard to analyze the topology of the molecules because it is hard to tell which of the strands at the points of their intersection goes above and which below the bed. To a large extent, this difficulty has been overcome for the first time by binding the double helix to the *recA* protein.[146] So much does the DNA strand grow thicker in the process, that the structure of the intersections of DNA segments becomes clearly visible in the photographs (Figure 50). On the other hand, knotted and linked DNA molecules differ in terms of their mobility in gel from unknotted and nonlinked molecules, which makes it possible to separate them in the process of gel electrophoresis (see the review by Wasserman and Cozzarelli[147]). This method naturally requires special calibration, for it is impossible to say in advance what position a particular topological structure must occupy relative to the unknotted circular DNA form. A sufficiently large body of experiment results on the mobility of various topological structures has been accumulated (Figure 51). In particular, the study by Liu and co-workers,[66] where knotted double-stranded DNA molecules were observed for the first time, showed that

FIGURE 50.   Electron micrograph of a DNA molecule forming a 5_1 knot. Sample preparation involved covering the molecule with the *recA* protein in enhance chain thickness. Supplied by E. M. Shekhtman, University of California, Berkeley.

the mobility of trefoils corresponded to the mobility of an unknotted supercoiled DNA with $\Delta Lk = -3$. Naturally enough, in the study of knotted and linked DNA molecules by this method, they must have single-strand breaks, for otherwise mobility will also depend on the double helix's linking number. Two important advantages of the electrophoretic method consist of its simplicity compared to electronic microscopy and the possibility of measuring the share of particular structures. It is impossible, of course, to identify knotted DNA molecules by the electrophoretic method, for such molecules are mirror stereoisomers. Electron microscopy is the only possible method here. It has notably shown that the trefoils formed with the help of topoisomerase II are distributed in equal shares between the left and right stereoisomers.[146]

The investigations of topological structures formed by double-stranded circular DNA gained further momentum with the discovery of the ability of topoisomerases II to change the topology of such molecules (see Chapter 2, Section IV). For the first time, however, knotted DNA molecules were discovered in single-stranded circular DNA preparations after their treatment with topoisomerase I.[148] That was the first case of the discovery of knotted polymeric chains.

Until recently, the question of the possibility of the existence of knotted DNA in living cells remained open. For the first time, knotted DNA molecules were separated out of *E. coli* cells.[149] It was shown that in a natural strain of such cells about 1% of pBR322 plasmid DNA precipitates out in the form of the 3_1 knot. In the strains carrying DNA-gyrase-mutated genes, the share of knotted DNA goes up to 10%. Most probably, the knots obstruct the normal functioning of DNA in the cell, although a certain share of knotted molecules

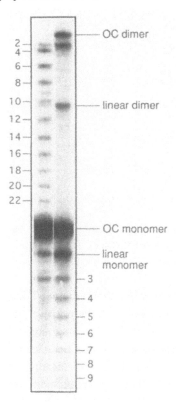

FIGURE 51. Electrophoretic separation of knotted (right lane) and linked (left lane) DNA molecules 4363 bp length. Each band corresponds knots or links with a specific minimum number of intersections in their projection. These numbers are shown next to each band. All links belong to the torus type of catenanes. OC dimer is an open circular DNA molecule of double length. Supplied by E. M. Shekhtman and D. E. Adams.

are formed in the process of DNA replication, whereas one of the functions of topoisomerases is to "untie" such knots. That is why the share of knotted molecules in DNA-gyrase-mutated cells is higher. Formerly, knotted DNA molecules had been isolated from the p4 phage,[150] but in that case the formation of knots apparently happens at the stage of the precipitation of phage DNA out of capsids. The point is that the DNA of the p4 phage has elongated cohesive ends and forms rings only at the stage of precipitation out of capsids, while its stacking in capsids may contribute to knotting.[150]

Unique opportunities for obtaining and studying the physical properties of various topological structures are offered by the use of special enzymic systems directionally changing the topology of DNA. An elegant example of such a study is the investigation of DNA supercoiling induced by the linkage of two molecules.[151] In that study, catenanes with various linking numbers were

FIGURE 52.   Diagram of torus catenanes. The linking number of the double-stranded
DNA rings is 7.

obtained by way of site-specific recombination. These catenanes belonged to
a special class of torus linkages, with the positive linking sign corresponding
to a right-handed helix (Figure 52). Single-stranded breaks were introduced
into those catenanes (approximately one per molecule) so that each of the
linked DNA should have an equilibrium distribution of twist and writhe. After
that, the integrity of the chains in the break spots was restored with the help
of DNA ligase. These experiments strongly resemble the assays with isolated
circular DNA described in Chapter 2, Section III. In this case, however, it was
natural to expect that the linkages would induce a non-zero mean writhe of
linked DNA. In principle, linkages also can induce changes in DNA twist if
the equilibrium twist changes at comparatively minor bends in the double helix
axis. In any case, after the restoration of the chains' integrity and the subsequent
cleavage of the double helix in one of the linked molecules, the second DNA,
which is free by now, ends up supercoiled. This supercoiling was measured
in quantitative terms depending on the linking number, lk, in the catenanes
(Figure 53). Note that the interpretation of the data suggested by Wasserman
et al.[151] was based on the mechanistic notion of the conformation of linked
DNA, corresponding to Figure 52. As revealed by accurate statistical-mechanical
simulation of this system,[152] the real conformations of linked molecules have
little resemblance to regular ones (Figure 54). The writhe of these molecules
undergoes great fluctuations, but its mean value grows in linear proportion to
the growth of lk. The dependence of induced writhe on lk, calculated through
simulation, very accurately coincided with the dependence of ΔLk on lk (Figure
53). This concurrence shows that the equilibrium twist of DNA undergoes no
changes in this experiment.

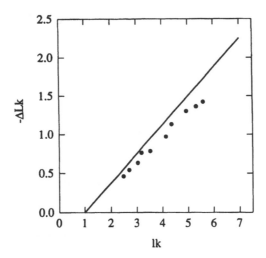

FIGURE 53. DNA supercoiling induced by catenation vs. the linking number, *lk*. Along with experimental data[151] (•), we present the results of theoretical simulation[152] (—).

FIGURE 54. Typical conformations of catenanes with a linking number of 7 (stereo pair). Simulation corresponds to open circular DNA molecules of 3.5 bp.[152]

Although there are ample experimental opportunities now for studying the statistical regularities of the formation of different topological structures by DNA molecules, this avenue of research remains practically unexplored. The only study on this subject was done by Wang and Schwarz[153] in 1967. They investigated the formation of linkages by DNA molecules of phages λ and 186 upon their spontaneous cyclization in solution. Because of the great length of these DNA, quantitative analysis of the results of this study in the light of the theoretical calculations outlined in the previous paragraph is unfortunately

FIGURE 55.   Diagram of site-specific recombination between two double-stranded DNA segments.

quite difficult. Similar experiments with shorter DNA and, especially, studies of the statistical knotting of DNA could provide a lot of information on the properties of the double helix.

D. Topological Approach to the Study of the Mechanisms of Enzymatic Reactions

Many enzymatic processes change the topological states of circular DNA molecules involved therein. One class of such reactions induced by topoisomerases was described in Chapter 2, Section IV. Another, more involved, class of processes, which also is the subject of intensive studies, is site-specific recombination. In general, the elementary act of such a recombination boils down to the introduction of two breaks in specific spatially close regions of DNA and a cross-wise restoration of chains (Figure 55). The topological state of the molecules may change upon the replication of DNA or in the course of the doubling of genetic material. A very effective method of studying such mechanisms has been worked out at the laboratory of N. Cozzarelli, University of California, Berkeley. It is based on the analysis of topological changes in circular DNA involved in these processes.[147,154] The elegant study which laid the foundation for this approach and which has to do with the working mechanism of DNA gyrase[57] was discussed in Chapter 2, Section IV. Here we shall examine another graphic example of the said approach, which involves the mechanism of site-specific recombination within the system of the enzyme resolvase, which is responsible for transporting transposons of the Tn3 family in the genome.

To study the mechanism of the action of resolvase, a special plasmid was constructed which carried two recombination sites specific to this system. Both sites were oriented in the plasmid in the same direction. Without retracing the whole path which led Wasserman et al.[155] to the formulation of the structural model of the recombination complex and of its working mechanism, let us simply follow the verification of this model. The model is shown schematically in Figure 56. According to the model, part of the complex shown in the stippled rectangle ought to have a fixed structure. It comprises the enzymes involved in the recombination, a DNA region with "three supercoils", and drawn-together recombination sites in a parallel orientation. Each recombination act leads to an additional "linkage" of regions belonging to different halves of the

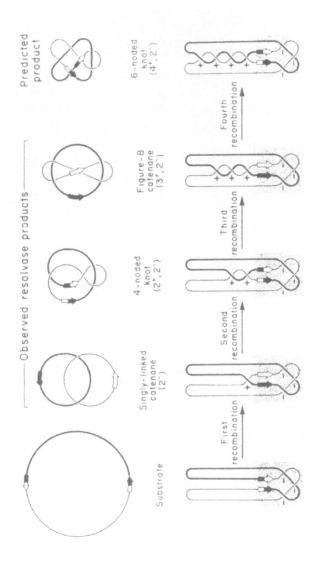

FIGURE 56. Diagram of site-specific recombination by resolvase Tn3.[155] The structure of the shaded part of the complex must be unchangeable. Specific recombination sites are indicated by arrows. In the top line, product topology of succesive rounds of recombination is presented in standard form.

molecule (separated by the recombination sites). According to this scheme, a single recombination act results in the formation of two circular molecules forming an elementary 2_1 linkage. This main product of the reaction is easily detectable in the system. As a result of the second recombination act, a circular molecule forming a 4_1 knot should evolve once again. The third recombination act generates two circles which now form a 5_1 linkage. All these products were spotted in the system, if relatively infrequently. The model in question was based precisely on the emergence of these products, whereas the verification of the model was based on the prediction of the formation of a 6_2 knot (see Figure 44) as the product of a quadruple recombination act. Such knots were indeed found in that system. To this end, the reaction products were separated by way of gel electrophoresis, whereupon DNA was isolated from the weak band corresponding to knots with six intersections in the standard form. This DNA was analyzed with an electron microscope. It turned out, in full conformity with the prediction, that all knots isolated in the model belonged to the 6_2 type. This effectively confirmed the validity of the suggested model of the recombination complex for this system.

The method of analysis for the topological state of the reaction products led to a better understanding of the site-specific recombination mechanisms in the Int phage λ,[156] Gin phage Mu,[157] and Hin[158] systems. On the other hand, the method makes it possible to procure information not only on the performance of specific enzymic systems, but also on the properties of the substrate itself, i.e., superhelical DNA. In particular, the set of knots and catenanes obtained from a specially constructed superhelical plasmid in the λ Int system showed that the superhelix had to have the interwound shape.[156] It was found that the topology of products obtained in this system depended on the degree of supercoiling in the original DNA. This made it possible to use the λ Int system for determining the number of turns in a twisted superhelix and for evaluating the torsional stress of DNA in a cell (see Chapter 6, Section I).

Chapter 4

FORMATION OF NONCANONICAL STRUCTURES UNDER THE IMPACT OF NEGATIVE SUPERCOILING

In the previous chapter we examined only those properties of circular DNA which are not related to disruptions of the regular structure of the double helix. However, with sufficiently high negative supercoiling, such disruptions become inevitable. In effect, such disruptions are one of the ways of its structural realization. In contrast to the alteration of twist and spatial curvature of the double helix axis, which are distributed comparatively evenly along the whole circular DNA molecule, disruptions in the regular structure have a local character, while their type and location depend on the DNA sequence. All these alternative structures are defined by a lesser (compared with the B-form) twist of one strand of the DNA relative to the other. In this chapter, we shall examine all known types of such structures and will carry out, wherever possible, qualitative and quantitative analysis of their occurrence under the impact of negative supercoiling. Recently, alternative structures have been studied with an ever-growing intensity. This is due not only to the fact that each such system constitutes a fascinating physical object, but also to a possible biological role of such structures. The question of the biological role of alternative structures in DNA is not very clear yet, and we are not going to examine it in this chapter. Some aspects of this problem are tackled in Chapter 6.

To end the introduction to this chapter, let us list the known types of alternative DNA structures arising under the impact of negative supercoiling.

The simplest types of structures induced by negative supercoiling are melted or open regions. In such regions, the strands of DNA are not interwoven around each other and may be actually twisted in the opposite direction. Such regions are bound to arise first of all in the DNA sections enriched with AT base pairs, because the melting of such pairs requires less free energy than does the melting of GC pairs.

In palindromic regions of DNA, cruciform structures may occur under the impact of negative supercoiling. One such structure is schematically presented in Figure 10. In the DNA section, which has assumed a cruciform structure, the complementary strands of DNA are not twisted relative to each other. At the same time, a large part of the palindromic region's nucleotides are involved in the helical structure of the cross's hairpins, which means that their free energy corresponds to free energy in a regular linear double helix. That is why the formation of a long enough cruciform structure must be preferable to the formation of open sections of the same size.

89

The maximum release of negative superhelical stress per base pair causes the formation of the Z-form, where the strands are twisted into a left-handed helix with a helical repeat of 12 bp per turn (see Chapter 1, Section I.C). The left-handed form predominantly occurs in DNA regions with a regular alternation of purines and pyrimidines.

Another type of noncanonical structure occasioned by negative supercoiling is the H-form, which may occur in DNA sections where one strand comprises only purine and the other only pyrimidine bases. The main element of the H-form is a triple helix. Topologically, the H-form is equivalent to an open section or a cruciform structure; the complementary strands of DNA in the H-form are not twisted relative to each other. The development of the H-form is stimulated by the reduction of pH of the solution.

Another structure where the double helix is somewhat unwound compared to the B-DNA form is A-DNA. The difference in the angles of helical rotation for these two forms is so insignificant, however, that negative supercoiling cannot by itself induce the transition of any section from the B- into the A-form. It can just slightly shift the B-A transition caused by some other factor, e.g., an increase in the concentration of alcohol in the solution.

It must be noted here that the formation of all the above noncanonical structures is thermodynamically wasteful in linear DNA in near-physiological conditions, and it is only negative supercoiling that allows them to evolve under such conditions.

I. EXPERIMENTAL METHODS OF ANALYSIS OF NONCANONICAL STRUCTURES IN CIRCULAR DNA

All the existing methods of studying noncanonical DNA structures occasioned by negative supercoiling can be divided into two groups. The first group brings together the methods based on the registration of local changes in DNA sections where a structural modification has taken place. In principle, these methods can be used just as effectively to register local structural changes in linear molecules also. We shall refer to this group as methods of localization of structural transitions. The second group of methods is based on the measurement of integral properties of DNA. As it is, only one method in this group, which is based on the use of the specific properties of closed circular DNA, is truly effective. On the formation of a noncanonical structure under the impact of negative supercoiling, superhelical stress itself decreases and so does the absolute writhe value which determines the electrophoretic mobility of DNA molecules in gel. This change in mobility at conformational transition lies at the core of the second approach. The method does not provide any information on where exactly the conformational change has happened, but allows the experimenter to obtain quantitative characteristics of the transition observed. The method offers an absolutely unique opportunity of reliable quantitative registration of conformational changes in a region that accounts for about 1% of the total length of the molecule. No other physical method registering integral properties of DNA (UV spectroscopy, circular dichroism, infrared spectroscopy, NMR) makes it possible to monitor changes affecting such a small fraction of DNA base pairs.

A. Methods of Localization of Structural Transitions

Most methods of localization of structural transitions are based on the application of breaks in DNA at the site of the formation of a noncanonical structure with the subsequent identification of the location of these breaks. Such breaks can be applied with the help of a special class of enzymes, endonucleases, which specifically hydrolyze single-stranded DNA. These endonucleases are sensitive even to such minor irregularities in the helical structure as, say, boundaries between B- and Z-helices.[159] In cruciform structures, they cause breaks in the loops of the hairpins.[160,161] There are sites sensitive to these endonucleases in the areas of the development of the H-form[162-165] and, naturally, in unwound regions of DNA. In supercoiled DNAs, there are also other areas of increased sensitivity to such endonucleases, though the nature of structural changes in such areas is not clear yet.[166] Although after the application of the first single-strand break noncanonical structures in DNA soon disappear as a result of the relaxation of superhelical stress, the enzymes often go on to cut the second strand of the DNA, opposite the first break, for they are highly specific to such areas as well. That is why a considerable share of the molecules end up in the linear form after the treatment with endonucleases. There is also an endonuclease which specifically cleaves DNA at the foot of cruciform structures (endonuclease VII of the T4 phage).[167]

Another way of applying breaks in the areas of the formation of noncanonical structures is by using chemical agents which modify DNA bases at the spots which in the regular B-form are screened off by the double helix. Further chemical treatment of such a selectively modified DNA leads to a break up of the sugar-phosphate chains at the sites of primary modification. The role of primary modifying agents is played by such compounds as diethyl pyrocarbonate,[168-172] which modifies unpaired purines, dimethyl sulfate,[168-172] reacting with guanines which have the N7 position unprotected, and osmium tetroxide,[168,173,174] which modifies unpaired thymines and others.[175,176] A whole series of reviews concerning the probing of noncanonical structures with various chemical agents and nucleases was published recently.[177] In the case of cruciform structures, modification happens first of all in the loops of the hairpins.[170,171,173-176] For DNA regions in the left-handed Z-form, the modification picture depends on the type of chemical agent and on the sequence of that region.[168,169,174] For example, in a region with a regular sequence $d(GC)_n$, which has assumed the Z-form, diethylpyrocarbonate reacts with all guanines of that region, while hydroxylamine reacts only with cytosines located at the junctions of the B- and Z-forms.[168,169] In the case of irregular sequences which are in the Z-form, the picture is more involved.[169] The method of chemical modification has proved very effective also for ascertaining the structure of the H-form of DNA.[178-180]

Both nucleases and chemical agents have been used for a comparatively long time for exploring the structure of supercoiled DNA.[181-183] Most probably, the formation of noncanonical structures would have been discovered much earlier had there been effective and accurate methods of localizing breaks resulting from nuclease and chemical treatment of DNA in the early 1970s. Until the mid-1970s, however, there were no such methods. They started to evolve only after the discovery of restriction endonucleases — enzymes strictly cleaving DNA at sites with certain base sequences. The discovery of these enzymes led to a genuine revolution in DNA research. Suffice it to mention the development of the methods of DNA sequencing on this basis. In its most popular version, the sequencing method boils down to the same task of accurately mapping specific breaks in DNA.[184]

In the cases when the area (or areas) of the occurrence of specific breaks are unknown, restriction analysis is used for rough preliminary mapping. DNA molecules converted into the linear form, by way of nuclease or chemical cutting, are treated with some restriction endonuclease to obtain a set of restriction fragments. This set is compared by way of electrophoretic separation by lengths with a reference set obtained as a result of the treatment of the original circular DNA with the same restriction endonuclease. If the cutting at the point of structural deformation occurred in one DNA region, two additional bands should appear in the electrophoregram of the first set of fragments, compared with the reference set. Besides, one of the lines of the calibration set should be missing in the first set of fragments (or become much less intensive, if the effectiveness of the linearization was incomplete). By gauging the length of the additional fragments according to the graduation of the gel and knowing localization within the sequence of the cut fragments, it is possible to find two possible sites of structural deformation. The results of such an experiment taken from the study[160] are presented in Figure 57. It is precisely the method whereby Lilley discovered the formation of cruciform structures in native DNA under the impact of negative supercoiling.[160] This approach allows one to localize the area of structural deformation with an accuracy of 10 to 50 bp.

After the preliminary localization of the site of specific cleavage of DNA, it is possible to obtain the picture of the distribution of cuts at the nucleotide

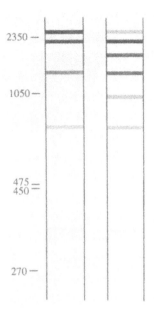

FIGURE 57. Diagrammatic electrophoretic patterns of fragments upon the treatment of supercoiled DNA with restriction endonuclease (left) and the same DNA pretreated with nuclease S1 (right). Shown on the left are the positions of reference fragments of known lengths.

level. This is done by a procedure used for DNA sequencing.[184] The only difference is that in the case of sequencing it is necessary to introduce breaks into a given DNA restriction fragment at certain types of bases, whereas with the mapping of chemical modification such breaks are applied by the modified sites of structural deformation. With nuclease probing of structural deformation, these breaks are applied by enzymes themselves. Electrophoretic separation of labeled single-chain fragments helps determine their length down to one nucleotide, and thereby map the breaks or chemical modification sites relative to the point of the endonuclease-induced cleavage. An example of such analysis is shown in Figure 58.

After the discovery of cruciform structures by the above method of mapping nuclease-induced cleavage, some experimenters opined that this method registered comparatively rare fluctuations of the DNA structure rather than its dominant states. In principle, such doubts were not unfounded. However, the follow-up studies and experiments with the use of different methods showed that with the appropriate selection of the experiment conditions the methods of nuclease-induced cleavage and chemical modification could produce accurate evaluations of the probability of the formation of noncanonical structures.[185]

Recently, experimenters have started using the photofootprinting method for the localization of the H-form of DNA.[186] This method is based on the

FIGURE 58. Autoadiographs of sequencing gel showing reaction of the d(CG)$_{16}$ repeat with diethylpyrocarbonate (DEP) and hydrozylamine (HA).[168] The vertical doudle-headed arrow spans the GC repeat. DEP shows enhanced reactivity with guanines in CG insert of supercoiled (SC) plasmid. HA shows high reactivity with cytosines in the region of the upper (5′) end of the CG repeat. Very weak reactivity in the region of the repeat was found for relaxed plasmid (line O for HA). Johnston, B. H. and Rich, A., *Cell*, 42, 713, 1985. With permission.

modification of the effectiveness of the formation of photoproducts in a DNA region upon its transition from the B-form to an alternative form. It has been found that the formation of a triple helix in the H-form of DNA practically blocks the formation of photoproducts (which are mostly pyrimidine dimers formed by neighboring pyrimidines in the DNA chain). Consequently, the method is very effective for this structure. Localizing the formation of photoproducts is based on the same principles of application of breaks at the sites of photolesions and subsequent localization of these breaks.

Another probing method is based on changes in the sensitivity to restriction endonucleases of DNA sections where noncanonical structures have evolved.[187] This method, of course, is applicable only when such regions contain specific recognition sites for the endonucleases used. The indisputable merit of this method is its exceptional simplicity.

In 1981, it was discovered that antibodies may arise for Z-DNA, which bond only with this conformation of the double helix.[188-191] This binding of anti-Z antibodies to supercoiled DNA was used in a number of studies as a method of registering the Z-form. Some of the experimenters mapped the binding sites of anti-Z antibodies on DNA.[192-194] As shown in the study,[195] however, the binding of anti-Z antibodies upsets the equilibrium of the B-Z transition in the direction of the formation of the Z-form. That is why the results obtained by this method must be treated with caution.

The most direct method of registering structural changes in DNA would be their electron microscopic observation. Unfortunately, such observation is rarely possible because of the method's insufficient resolution. The only exception is the observation of big cruciform structures formed in palindromic areas of DNA out of hundreds or thousands of base pairs. This is precisely the method whereby the formation of cruciform structures in a recombinant plasmid, which is essentially a couple of DNA molecules knitted together, was observed for the first time in the study.[196]

B. The Method of Two-Dimensional Gel Electrophoresis

This very elegant and effective method of analyzing local conformational changes occasioned by negative supercoiling was first suggested by Wang et al.[197] The method allows one to obtain, as a result of a single experiment, the complete dependence of the probability of the formation of an alternative structure on the superhelical density. Besides, it yields important information on the changing equilibrium number of turns of one DNA strand relative to the other, which is associated with this transition. At the moment, two-dimensional electrophoresis is the principal method of quantitative analysis of conformational transitions in circular DNA.

We shall begin the description of the method by analyzing how DNA topoisomers will move in an ordinary one-dimensional gel. The mobility of molecules in this kind of electrophoresis monotonously increases with the growth of the absolute value linking number difference. This is due to the fact that topoisomers with a larger linking number difference must have a greater writhe and, consequently, a more compact conformation. This, however, is true only as long as the molecule undergoes no conformational changes which reduce superhelical stress and, consequently, the absolute writhe value. Topoisomers which have undergone the transition will have lower mobility compared to what it would have been in the absence of conformational changes. As a result, the monotonous rise in the mobility of molecules with the growing linking number difference may be disrupted. An electrophoregram of the mixture of topoisomers will show a pattern of irregular lines by which it will be hard to identify specific topoisomers (Figure 59).

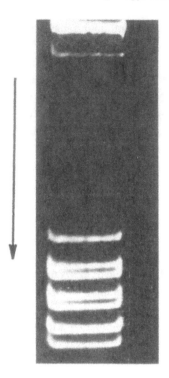

FIGURE 59. One-dimensional separation of pAO3 DNA topoisomers. Some bands in the lower part of the pattern correspond to topoisomers in which a structural transition has occurred, so it is no longer possible to unambiguously assign the bands to specific topoisomers.

It is precisely to correlate the lines in such a "chaotic" electrophoregram with specific topoisomers that two-dimensional electrophoresis is used (the use of this method for the determination of topoisomer distribution in a DNA preparation was described in Chapter 2, Section II.A). Electrophoresis in the first direction is effected under ordinary conditions, along one of the edges of a flat gel. This results in the distribution of topoisomers which was described above. Before the electrophoresis in the second direction, the gel is saturated with a ligand intercalating between the DNA bases. As a result of the ligand binding, the superhelical stress in DNA is no longer determined by the linking number difference, ΔLk, but by the value ($\Delta Lk - v\phi N/360$), where v is the number of bound ligand molecules per base pair, ϕ is the angle of unwinding of the double helix upon the binding of a single ligand molecule ($\phi \leq 0$), and N is the number of base pairs in DNA. The concentration of the ligand is selected in such a way that the remaining superhelical stress should be insufficient for the formation of a noncanonical structure even in a single topoisomer. In this case, the mobility of topoisomers will monotonously depend on the absolute

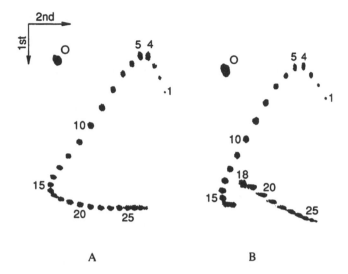

FIGURE 60. Two-dimensional gel electrophoresis of pBR322 DNA (A) and of the same DNA with a $d(CG)_{16} \cdot d(CG)_{16}$ insert (B). (From Wang, J. C., Peck, L. J., and Becherer, K., *Cold Spring Harbor Symp. Quant. Biol.*, 47, 85, 1983. With permission.) Upon electrophoresis in the vertical direction, the gel was saturated with chloroquine, which caused a partial relaxation of superhelical stress, and the 15th topoisomer (which had 10.5 superlcoils in the first-direction electrophoresis) came to have the minimum mobility. The structural transition in topoisomers from number 18 onwards in the insert-carrying DNA manifested itself as a leap of mobility in the vertical-direction electrophoresis.

value of the sum ($\Delta Lk - v\phi N/360$). That is why after the completion of electrophoresis in the second direction perpendicular to the first one, correlating spots with specific topoisomers will not be difficult at all. The abrupt reduction of the mobility of topoisomers as a result of a cooperative transition of a DNA region into a noncanonical structure, which occurred in the course of the electrophoresis in the first direction, will be very manifest indeed.

Let us examine as an example an electrophoregram of a mixture of topoisomers of pBR322 DNA and of the same DNA with a $d(CG)_{16} \cdot d(CG)_{16}$ insert, transforming into a noncanonical structure under the impact of negative supercoiling.[197] As follows from Figure 60A, the mobility of topoisomers in the course of vertical electrophoresis monotonously rises (starting from the 5th, which corresponds to $\Delta Lk \cong 0.5$) and gradually reaches saturation in the vicinity of the 20th. Topoisomers with the 18th and higher numbers can only be separated during horizontal electrophoresis which was conducted after the saturation of the gel with the chloroquine ligand intercalating in DNA. In this case, the mobility minimum shifts onto the 15th topoisomer, and all topoisomers become easily separable. The electrophoregram of the same DNA with an insert shows a different picture (Figure 60B). The mobility of topoisomers of this DNA in the course of vertical electrophoresis leaps sharply between the 17th and 18th, which becomes clear after the completion of electrophoresis in the second, horizontal direction. This mobility leap is apparently associated with the formation of some noncanonical structure in the insert, which attenuates superhelical stress and reduces the absolute value of writhe and, consequently, of the DNA's mobility. This may be a Z-form or a cruciform structure, because the insert is formed by a sequence which easily assumes the Z-form and is also palindromic.

Apart from the superhelix density interval in which a DNA region turns into a noncanonical structure, one can procure another very important piece of information from such an electrophoregram. It is possible to determine for each topoisomer the mean change in the equilibrium twist of one strand around the other as a result of the transition, which is normally designated as δTw. Figure 60B shows that the mobility of the 20th topoisomer during the vertical electrophoresis becomes almost equivalent to that of the 14th. It means that these topoisomers must have the same writhe and, consequently, the same superhelical stress. The formation of a noncanonical structure must have only very slightly changed general torsional rigidity of the DNA, because the transition affected a very small share of its base pairs. This means that with equal superhelical stress the deviation in the strands' twist from the equilibrium value must be the same also. Consequently, the whole difference in the linking number in these topoisomers is compensated with the change of the equilibrium twist during the transition. In particular, for the 20th topoisomer ($\Delta Lk = -15.7$) the mean decrease in the equilibrium twist, δTw, is approximately 5.7 turns (its mobility corresponds to a "topoisomer with the number 14.3"). Figure 61 shows the dependence of δTw on ΔLk, built on the basis of the electrophoregram under review. When plotting this dependence, we assumed that the mobility of all topoisomers up to the 16th is not affected by the transition. The validity of this supposition is confirmed by a comparison of the two electrophoregrams shown in Figure 60. That is why the positions of the topoisomers 13 to 16 on the electrophoregram can be used as reference points for determining δTw for the topoisomers 17 to 21. Figure 61 shows that the total change of δTw as a result of the transition is equivalent to 5.7 turns. This unequivocally indicates that in the course of the experiment the $d(CG)_{16} \cdot d(CG)_{16}$ insert assumes a left-handed Z-form rather than a cruciform. Indeed, with a B-Z transition the total change in the twist corresponds to the unwinding of 32 bp in the right-handed helix (with a helical repeat 10.5 bp per turn) and their coiling into a left-handed helix (with a helical repeat 12 bp per turn). This means that the change in the twist δTw must be $32/10.5 + 32/12 = 5.7$ turns, which is in full conformity with the experiment. In the case of the formation

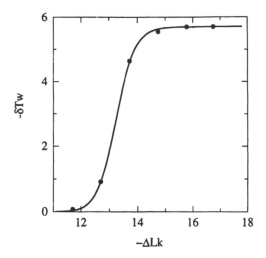

FIGURE 61. The changing equilibrium twist of DNA owing to the transition of the $d(CG)_{16} \cdot d(CG)_{16}$ to the Z-form vs. the linking number difference. This plot was constructed on the basis of the data presented in Figure 60.

of a cruciform structure, δTw corresponds only to the unwinding of the right-handed helix and must be equivalent to $32/10.5 \cong 3$ turns.

It must be noted, however, that the actually observed δTw value far from always correlates so well with the expected one. In particular, for cruciform structures the observed δTw value is often 1 to 1.5 turns larger than the estimated value.[198,199] This is due, at least partially, to the deceleration of DNA molecules in gel as a result of the formation of the cruciform structure itself. This is confirmed by the findings of the study[200] which explored the movement of linear DNA fragments with stable cruciform structures that cannot transform into a regular double helix (see Figure 9). There is, probably, some additional effect from the emergence of borderlines between the B-form and the newly formed noncanonical structure. In accordance with the findings of the study,[201] the structure of these borders changes in response to changes in superhelical stress, which explains the smooth growth of δTw beyond the transition area, which is often observed in experiments. Even so, in the above example, the conclusion about the transition of the insert into the Z-form is beyond any doubt.

The δTw value observed in the experiment is the mean value of the twist change in the course of electrophoresis for the given topoisomer. That is why if the period of relaxation for the formation of a noncanonical structure exceeds the period of electrophoresis, some topoisomers may appear in the electrophoregram in the form of two spots indicating the presence or absence of a noncanonical structure. This situation can really be observed in the case of cruciform structures (see Figure 63) whose relaxation period can run into dozens of hours.[198]

As follows from Figure 60, for the interpretation of two-dimensional electrophoresis experiments, the DNA preparation must contain a broad set of topoisomers comprising both molecules with $\Delta Lk \geq 0$ and molecules in which a noncanonical structure has evolved. Such a set can be obtained by treating circular DNA with topoisomerase I with different concentrations of the ligand intercalating in DNA (see Chapter 2, Section II.D).

The two-dimensional electrophoresis method has one limitation which is absent from the positional methods of investigation of noncanonical structures. For the leap in mobility to be visible in the electrophoregram, the change in the mobility of topoisomers in the conformational transition range during vertical electrophoresis must not reach the saturation point. This means that the transition must happen with a not too high superhelical density of the supercoils.

II. THERMODYNAMIC ANALYSIS OF THE FORMATION OF NONCANONICAL STRUCTURES

In this section we shall present a basic thermodynamic analysis of the formation of noncanonical structures in closed circular DNA. This analysis leaves out the important fact that in real DNA a multitude of close states may occur with a comparable probability, so it is not always suitable for quantitative analysis of the experimental data. This is better done by means of the statistical-mechanical approach which will be explained in Section VIII. Even so, thermodynamic analysis is very useful for comprehending the main regularities of the formation of noncanonical structures in circular DNA, while the relationships derived from this analysis are very simple in form and have a clear physical meaning.

Let us first of all note that all analysis in this chapter is based on the generally accepted assumption that the free energy of the formation of noncanonical structures can be presented in the form of two summands. One of these summands corresponds to the change in the free energy of linear DNA upon the conformational change, and the second to the transition-related change in the free energy of supercoiling. Let us start with the analysis of the second summand. As repeatedly noted above, the supercoiling of circular DNA molecules leaves them with additional free energy which is known as the energy of supercoiling. In a first approximation, this free energy proves proportional to the square of the linking number difference in the DNA [see Chapter 2, Equation (2.15)]. In the case when the structure of the DNA is a regular double helix in the B-form, the linking number difference characterizes the molecule's elastic deformation, which means that the energy of supercoiling is simply proportional to the linking number difference squared, and, in this sense, there is an analogy with Hooke's law. How will the energy of supercoiling change if m links of the molecule transform into a noncanonical structure? The natural assumption is that the energy of supercoiling will continue to depend only on the DNA's elastic deformation. The magnitude of elastic deformation, however, will now be decided not by the linking number difference ΔLk, but by a different $(\Delta Lk - \delta Tw)$ value, where δTw stands for the change in the equilibrium twist at the transition of base pairs from the B-form to the alternative structure under review. The δTw value can easily be expressed through the number of base pairs γ_B per turn of the double helix in the B-form and the number of base pairs per turn of one strand around the other in the alternative structure γ_{alt}:

$$\delta Tw = m\left(1 / \gamma_B - 1 / \gamma_{alt}\right) \qquad (4.1)$$

We shall henceforth mark the $(1 - \gamma_B/\gamma_{alt})$ value as κ. Thereby, the energy of supercoiling can now be expressed in the form [see Chapter 2, Equation (2.15)]:

$$G = KRT / N \cdot \left(\Delta Lk + m\kappa / \gamma_B\right)^2 \qquad (4.2)$$

or by expressing it through the superhelical density:

$$G = AN(\sigma + m\kappa / N)^2 \qquad (4.3)$$

A and K in these equations stand for numerical coefficients $(A = KRT/\gamma_B^2)$ and N for the number of base pairs in the molecule [see the relationships in Chapter 2, Equations (2.15) and (2.16)]. Accordingly, the change in the free energy of supercoiling ΔG at the transition of m base pairs into the alternative structure under review equals:

$$\Delta G = AN\left[(\sigma + m\kappa / N)^2 - \sigma^2\right] \qquad (4.4)$$

This equation formally implies that the formation of any structure with a lesser twist of the strands than in the B-form of DNA reduces the energy of supercoiling (for a negative σ) and, consequently, must be stimulated with negative supercoiling. This, of course, is true only under the condition that this structure forms in a comparatively short DNA fragment, so that the

$m\kappa/N$ value does not exceed the absolute value of the superhelical density σ. Conversely, with $m\kappa/N \geq -\sigma$ the expansion of the structure under review will enhance the energy of supercoiling.

It is not hard to generalize Equation (4.4) for the event of the formation of several local noncanonical structures, each of which is characterized by its own m_i and κ_i values:

$$\Delta G = AN\left[\left(\sigma + \sum_i m_i\kappa_i / N\right)^2 - \sigma^2\right] \tag{4.5}$$

Equation (4.5) leads to a very important conclusion. The change in the free energy of supercoiling at the formation of this noncanonical structure must depend not only on the superhelical density in a circular DNA, but also on what other noncanonical structures exist in that molecule. In other words, there is mutual influence between different conformational transitions in closed circular DNA, which does not depend on the distance along the chain between the rival structures. In this sense, closed circular DNA differs radically from linear molecules. It is obvious that because of this influence it is actually necessary to take into consideration in theoretical analysis the possibility of the formation of all permissible noncanonical structures in this sequence. All too often, however, we have to deal with a region where a particular transition in a particular section of the sequence happens under conditions where no other transitions are in evidence in either that region or other parts of the molecule. Analysis of such situations makes it possible to ascertain a number of general regularities about structural transitions under the impact of negative supercoiling. That is why we shall start with the examination of this situation.

We shall examine a circular DNA consisting of N base pairs and containing a section of n pairs capable of forming a certain noncanonical structure (this analysis was first carried out for the B-Z transition[202]). Let us designate this structure as κ, free energy of a base pair's transition from the B-form to the alternative structure as ΔF, and free energy of each boundary as F_j. Let us assume that ΔF is equal for all pairs in the region. A cooperative transition in this region precipitated by the increase in the superhelical density does not necessarily encompass the whole section, but may include a shorter stretch of m base pairs. The full free energy change upon the transition of a stretch of base pairs into the structure under review equals:

$$\Phi = \Delta G + m\Delta F + 2F_j \tag{4.6}$$

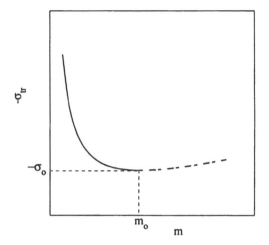

FIGURE 62.   Dependence of the superhelix density at which a noncanonical structure is formed on its size (the number of base pairs in the segment involved). The curve corresponds to Equation (4.7). The dashed part of the curve is not realized in the experiment, as the cooperative transition in this case, for $n \geq m_0$, involves only m_0 base pairs.

At the transition point, σ_{tr}, the full free energy change Φ must be zero. That is why, on the basis of Equations (4.4) and (4.6), we get the equation for σ_{tr}:

$$-\sigma_{tr} = m\kappa / (2N) + \Delta F / (2A\kappa) + F_j / (Am\kappa) \qquad (4.7)$$

The dependence of σ_{tr} on m, corresponding to Equation (4.7), is shown in Figure 62. The minimum value $(-\sigma_{tr})$ is achieved at the m_0 point. The coordinates of this minimum can be derived from the following equations:

$$m_0 = \frac{1}{\kappa}\sqrt{2NF_j / A} \qquad (4.8)$$

$$-\sigma_{tr} = \Delta F / (2A\kappa) + \sqrt{2F_j / (AN)} \qquad (4.9)$$

Let as assume that we gradually raise $(-\sigma)$ for the given circular DNA. It is clear that the cooperative transition must occur at the minimum possible $(-\sigma_{tr})$ for this section. Depending on the proportion between m_0 and n, we should observe two essentially different types of behavior of our system. With $n \leq m_0$, the minimum $(-\sigma_{tr})$ occurs at $m = n$, and the cooperative transition encompasses the whole region of base pairs capable of forming the alternative structure. The transition point in this case will be decided by Equation (4.7) in which m must be replaced with n.

With $n \geq m_0$, the minimum $(-\sigma_{tr})$ equals $(-\sigma_{tr}^0)$, and the cooperative transition will involve only m_0 base pairs. In this case, to determine the transition point, we must replace m in Equation (4.7) with m_0 [or use the Equation (4.9)]. Such behavior results from the release of superhelical stress at transition, which is the primary cause of structural rearrangement in this case. The stress remaining after the transition of m_0 pairs is insufficient for the expansion of the size of the noncanonical structure. It is only with further growth of $|\sigma|$ that the newly formed noncanonical structure will expand. The gain in supercoiling energy at the addition of one pair to the newly formed structure, which amounts to $2A\kappa(\sigma + m\kappa/N)$, must compensate for the increased free energy ΔF of that pair, resulting from its transition to an alternative structure, which means that the following condition must be met:

$$\Delta F + 2A\kappa(\sigma + m\kappa / N) = 0 \qquad (4.10)$$

In other words, with the expansion of the size of the structure from m_0 to n, the $(\sigma + m\kappa/N)$ value must remain unchanged. It is precisely these values, incidentally, that determines the molecules' mobility in gel, which means that during the expansion of the size of the structure their mobility must not change.

For a cooperative transition of n pairs into the alternative structure under review, or in the case of $n \leq m_0$, the notions of the degree of transition ϑ and the width of transition $\Delta\sigma$ may be introduced. The degree of transition corresponds to the probability of a region being in the alternative form for a given σ and is governed by the following equation:

$$\vartheta = \exp(-\Phi / RT) / \left[1 + \exp(-\Phi / RT)\right] \qquad (4.11)$$

where Φ must be viewed as a function of σ and calculated in accordance with Equation (4.6). The width of the transition is determined from the value of the derivative $(d\vartheta/d\sigma)$, calculated at the point $(-\sigma_{tr})$:

$$\Delta\sigma = \left|1 / (d\vartheta / d\sigma)\right| \qquad (4.12)$$

By using Equations. (4.6) and (4.11), one can easily arrive at the following expression for $\Delta\sigma$:

$$\Delta\sigma = \frac{2RT}{An\kappa} \qquad (4.13)$$

It should be noted here that Equation (4.13) is based on the square dependence of the free energy of supercoiling on the superhelical density, $G(\sigma)$. In the more general case, the constant A in Equation (4.13) should be replaced with the $[d^2G/d\sigma]/(2N)$ value. Quantitative analysis shows that even with comparatively minor deviations of the $G(\sigma)$ from the square dependence, the width of the transition may change quite substantially.[63,203]

In the following sections, we shall examine the application of the above dependences to the analysis of specific structural transitions.

III. CRUCIFORM STRUCTURES

The formation of cruciform structures in native circular DNA was discovered at the end of 1980 by Lilley[160] and, almost at the same time, by Panayotatos and Wells.[161] Some information on these structures, however, had appeared much earlier. It was back in 1975 that Hsieh and Wang predicted the possibility of the formation of such structures on the basis of thermodynamic analysis.[59] At that time, however, it was unclear what size palindromic zones could occur in real native DNA; the first phage DNA was sequenced only in 1977.[204] In 1979, it was demonstrated on the basis of statistical-mechanical analysis[205] that a cruciform structure could form in one of the palindromes of that DNA in the physiological superhelical density range. Also in 1979, Gellert and co-workers demonstrated the formation of cruciform structures in a recombinant plasmid which is essentially a giant palindrome (this plasmid was obtained by head-to-head cross-linking of two ordinary plasmids).[196] Even so, hardly anyone expected at that time that cruciform structures would be found in practically all DNA with the corresponding superhelical density isolated from cells. In his pioneering work, Lilley studied the DNA of the ϕX174 phage and of the ColE1 and pBR322 plasmids and detected the formation of cruciform structures in all the three DNAs. The remarkable thing is that in the pBR322 DNA these structures formed with a different degree of effectiveness in three different palindromes.[160] The size of all palindromic areas where cruciform structures were registered proved almost identical. It corresponded to the formation of 9 to 13 bp in the helical regions of each of the cross' hairpins, while the hairpin loops consisted of 3 to 5 bases. Similar results were obtained in Wells' laboratory, University of Birmingham by the same method of nuclease probing.[161] Theoretical analysis[206] indicated that such cruciform structures must occur in other phage and plasmid DNAs also. That was when intensive research into cruciform structures developing under the impact of negative supercoiling began.

The primary task was to find out whether the endonuclease (see Section I) used for the registration of cruciform structures had any role in their formation and to calculate the quantitative probability of the formation of such structures

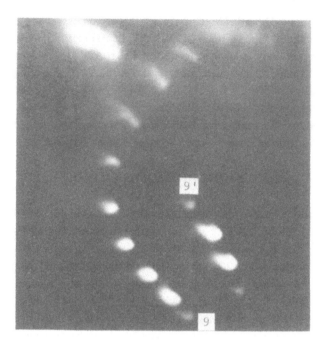

FIGURE 63.   Two-dimensional electrophoresis of pAO3 DNA.[197] The mobility leap between the 8th and 10th topoisomers corresponds to the formation of a cruciform structure in this DNA. The 9th topoisomer is present in two conformations, with linear (9) and cruciform (9′) states of the palindrome.

depending on the superhelical density. This task was first accomplished in the study[207] with a fairly elaborate use of the same nuclease method. This was shortly followed by the successful use of two-dimensional gel electrophoresis with regard to the formation of cruciform structures.[198] The electrophoretic pattern of the separation of topoisomers for the pAO3 plasmid[198] is shown in Figure 63. This DNA is a quarter of the ColE1 plasmid, comprising the latter's main palindrome. The figure shows that the transition happens between the 8th and 10th topoisomers, which corresponds to a supercoil density of about –0.06. It was discovered for the first time in that study that the period of relaxation at the formation of cruciform structures could be dozens of hours or even longer. Indeed, the 9th topoisomer shows in the electrophoregram as two spots corresponding to molecules with and without a cross (see Figure 63). This means that over the whole period of electrophoresis, which lasts for about 20 h, the conformational state of the palindromic zone in most of the molecules in that group did not change (for otherwise they would have ended up between the two extreme positions of this topoisomer on the electrophoregram).

With the advent of the first quantitative studies of the formation of cruciform structures it became possible to analyze to what extent that process matched the existing theoretical notions and to verify its energy parameters. Of the two energy parameters introduced in Section II for describing the energy processes attendant to the formation of an alternative structure in DNA, in the case of cruciforms the ΔF value must be assumed to be zero. Indeed, the cross grows in size at the expense of the destruction of 2 bp in the main helix and the formation of two identical base pairs in the hairpins (this process of the displacement of the crossing point is presented in Figure 82). This should not change the free energy of the structure, because the number of paired bases and all border elements of the structure remain as they were. This means that the formation of cruciform structures must be characterized by a single energy parameter F_j (only the $2F_j$ value has a physical meaning in this case, so this is precisely what we shall deal with below).

The main equation relating the transition point σ_{tr} of a palindrome of n links into a cruciform structure, and the free energy $2F_j$ of the formation of that structure looks as follows [see also Equation (4.7)]:

$$-\sigma_{tr} = F_j \,/\, An + n \,/\, (2N) \qquad (4.14)$$

In passing from Equations (4.7) to (4.14), we reckoned with the fact that for cruciform structures the parameter κ is equal to 1. Indeed, their formation is accompanied by the complete unwinding of the corresponding region of B-DNA, while in cruciform structures the strands are not twisted relative to each other [$\gamma_{alt} = \infty$, see Equation (4.1)]. The parameter A, according to the currently available data, is assumed to equal $10RT$ for all ionic conditions. Equation (4.14) describes a transition only if $n \leq m_0$, which means that the whole palindrome transforms into a cross as a whole. To evaluate m_0, we must know the parameter $2F_j$ [see Equation (4.8)].

Let us evaluate the parameter $2F_j$ on the basis of the experimental data presented in Figure 63. The electrophoretic pattern shows that the entire palindrome transforms into a cruciform structure as a whole; the entire change in the DNA's equilibrium twist, δTw, happens in a narrow interval of σ values. That is why the $2F_j$ can be calculated from Equation (4.14). The palindrome of DNA pAO3 in which the cruciform structure evolves consists of 31 bp, whereas the length of the whole DNA is 1683 bp.[198] By using Equation (4.14), we thus find (bearing in mind that $\sigma_{tr} = -0.06$) that $2F_j = 21$ kcal/mol. The obtained $2F_j$ value can be used for evaluating the maximum size of the palindromic zone which is yet to transform into a cross as a whole. By using Equation (4.8), we obtain:

$$m_o = \sqrt{2NF_j / A} \cong \sqrt{3N} \qquad (4.15)$$

It follows from this formula that in nearly all cases of natural palindromes whose length normally does not exceed 40 bp, the transition will be fully cooperative.

Unfortunately, the formation of cruciform structures cannot always be adequately described by Equation (4.14). In particular, for palindromes with $d(AT)_n$ sequences, the value of the $2F_j$ parameter, which can be determined from experimental data obtained for several palindromes of different lengths,[199] changes within 3 kcal/mol. This is probably due to the fact that in a number of cases, the formation of cruciform structures leads to additional unwinding of DNA compared to what was expected. Analysis of the studies[187,198,199,208-210] shows that for different palindromes forming cruciform structures with 3 to 5 bp in the hairpin loops, the parameter $2F_j$ varies from 15 to 21 kcal/mol. Because of these fluctuations in the $2F_j$ value, in the general case we can only predict the superhelical density at which cruciform structures are formed to within 0.01.

A number of studies dealt with the impact of external conditions on the formation of cruciform structures.[211,212] It is clear that as one approaches the melting point of linear DNA the probability of the formation of cruciform structures should decrease because of the formation of open regions, which should lead to a release of superhelical stress.[206] In the experiment, however, the disappearance of cruciform structures with rising temperature is observed considerably earlier than the expected emergence of open regions.[212] It is possible that some other structures join in the competition here whose formation is stimulated by higher temperatures. This conclusion is circumstantially confirmed by the fact that an extended palindrome, where the transition is observed at a low supercoil density ($\sigma_{tr} = -0.03$), forms a cruciform structure even at 65°C;[187] the formation of rival structures is less likely at such a low superhelical density.

Superhelical density at which the transition of palindromic sequences into cruciform structures takes place depends on the ionic conditions in the solution.[211,213] Figure 64 shows data for three palindromes obtained by way of two-dimensional gel electrophoresis.[213] It is clear from the figure that a decrease in the concentration of sodium ions in the solution leads to a considerable drop in the absolute values of σ_{tr} at which the transition takes place. The influence of the ionic conditions must be taken into consideration in any quantitative analysis associated with the formation of cruciform structures.

Totally unexpected results were obtained in the study[214] which explored the influence of Mg^{2+} ions on structural changes in the supercoiled DNA of SV40. The authors of this study found out by way of endonuclease probing that at

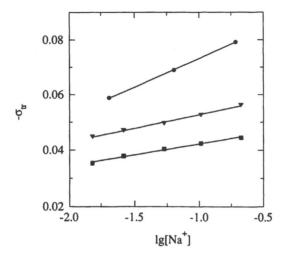

FIGURE 64.   Dependence of the superhelix density at which cruciform structures occur on ionic conditions.[213] We present data for the ColE1 DNA palindrome (Figure 101) (●) and for the inserts $d(AT)_{11}$ (▼) and $d(AT)_{21}$ (■).

a low concentration of Na^+ and 1 mM Mg^{2+}, two cruciform structures arose in that DNA in very short palindromes. The helical zones of the crosses' hairpins in this case had to consist of only 5 bp. If we use the known minimum value of the parameter $2F_j$, which is 15 kcal/mol, we can calculate from Equation (4.14) that $\sigma_{tr} = -0.09$ ($N = 5242$, $n = 13$). Under ordinary conditions, torsional stress matching such a high superhelical density never occurs in DNA because of the formation of noncanonical structures. In the study,[214] the superhelical density was not monitored, but for SV40 DNA isolated from a cell it stands at about -0.05.[47] That is why the results of this study sharply clash with the known regularities of the formation of cruciform structures, so an independent confirmation of these data would be in order before any conclusions can be made.

IV. THE LEFT-HANDED Z-FORM

A sensational discovery was made in 1979 in the laboratory of A. Rich, Massachusetts Institute of Technology, Cambridge. It was found that the $d(CG)_3$ hexanucleotide could exist in the form of a left-handed double helix which was called the Z-form.[33] This structure formed at the crystallization of the hexanucleotide from a solution with a very high salt concentration. It became clear right after that discovery, however, that the formation of Z-DNA could be stimulated not only by adding salt to the solution, but also by negative supercoiling. Indeed, the transition of a DNA fragment from a right-handed

B-helix into a left-handed Z-helix should precipitate the maximum release of superhelical stress compared to other noncanonical structures. And sure enough, comparatively soon it was proved that under the impact of negative supercoiling fragments with $d(CG)_n$ sequences could assume the Z-form under ordinary ionic conditions.[159,197] In fact, somewhat later it became clear that not only sequences with a regular alternation of purines and pyrimidines, as suggested in the study,[33] but also many others can be converted into the Z-form as long as the superhelical stress is high enough.[215,216] At present, we have a pretty full picture of the regularities governing the formation of the Z-form under the impact of negative supercoiling. Let us first have a qualitative description of this picture and then examine the quantitative regularities of the formation of the Z-form in DNA with a random base sequence. We assume that the reader is familiar with the basic data on Z-DNA outlined in Chapter 1, Sections I.C and III.E.

First of all, note an important methodological feature of the studies of the B-Z transition in circular DNA. In contrast to the situation with cruciform structures, the studies of the evolvement of the Z-form were based from the very start on the use of recombinant plasmids with deliberately integrated sequences liable to adopt the Z-form. The reason for this is that plasmid and phage DNAs which are convenient for this kind of research lack long enough fragments capable of transforming into the Z-form at a low enough superhelical density. It can definitely be asserted that the progress accomplished to date in the studies of the B-Z transition would not have been possible without the use of genetic engineering methods.

The B-Z transition happening under the effect of negative supercoiling in fragments with $d(CG)_n$ under ordinary ionic conditions was first discovered by Singleton et al.[159] They used the same method of nuclease probing, whereby the formation of cruciform structures in circular DNA had been discovered. It turned out that endonucleases specific to single-stranded DNA occasioned breaks at the junctions of the B- and Z-forms of the double helix, or on the borders of the inserts built into the DNA. Pretty soon, however, the nuclease probing method, even as other methods of Z-DNA registration, was ousted in the analysis of the transition in inserts by two-dimensional gel electrophoresis which was first suggested by Wang et al.[197] The first experiment to register the B-Z transition by this method was described in detail in Section I.B. Soon, the said transition caused by negative supercoiling was discovered in another purine-pyrimidine sequence, also $d(AC)_n \cdot d(GT)_n$.[217] It was demonstrated in a number of studies that the superhelical density at which the transition occurred in these inserts largely depended on ionic conditions in the solution. As the concentration of ions goes up, the $(-\sigma_{tr})$ value first increases also, but then starts to go down.[159,218-221] The drop in $(-\sigma_{tr})$ at a high concentration of Na^+ ions is associated with the decrease in the free energy of transition in linear

DNA, and the poly[d(CG)] polynucleotide assumes the Z-form for [Na$^+$] = 2.5 M.[219,221,222] The maximum magnitude of the free energy of transition is observed at [Na$^+$] = 0.2 M. Many bi- and trivalent ions are instrumental in the B-Z transition also (see the review by Rich et al.[4]).

A temperature rise slightly shifts the B-Z balance towards the stabilization of the Z-form. This has been confirmed by experiments with linear polynucleotides[223] and also inserts built into circular DNA.[224]

The d(CG)$_n$ · d(CG)$_n$ and d(AC)$_n$ · d(GT)$_n$ sequences have proved that only the regular ones in which the B-Z transition caused by negative supercoiling were observed. Little by little it has turned out, however, that there are no strict limitations for sequences capable of assuming the Z-form.

As a rule, the absolute value of the superhelical density in circular DNA does not exceed 0.1, although it can be raised to 0.2 with high concentrations of a ligand intercalating into DNA and of topoisomerase I. But can the superhelical density be −1? This value corresponds to closed but not linked single-stranded complementary rings. Such a system was first obtained in the studies by Brahms et al.[215] and Pohl et al.[216] Because of topological limitations, such rings cannot form a right-handed double helix throughout the length. The structure that such rings form in solution was designated the V-form. It has been demonstrated by different methods that about 30% of bases in the V-form make up a left-handed Z-helix and about as many a right-handed B-helix, whereas the remaining part of the DNA stays in a denatured state.[215,216] The sequence of the DNA under review did not contain an appreciable share of purine-pyrimidine blocks, which means that the Z-form arose in that DNA in fragments with irregular sequences also. This signified that given a sufficient torsional stress, fragments with a very broad range of sequences can be converted into the Z-form also. This conclusion was confirmed by X-ray analysis of the d(CGATCG) hexanucleotide which formed an unbroken Z-helix.[225] These studies raised the question of the the quantitative characteristics of the B-Z transition in DNA with a random base sequence.

This question can be addressed on the basis of a statistical-mechanical model of the B-Z transition in linear DNA. This model was first suggested in a study[34] and was described in Chapter 1, Section III.E of this book. The model contains six energy parameters which had to be determined by comparing theory and experiment. It was also necessary to ascertain the adequacy of the suggested theoretical description by checking the main assumption of the model, according to which the energy of a pair in the Z-form is determined in a good approximation only by the type of the pair itself and its conformations (I or II) and does not depend on the type or conformation of the nearest neighbors. Both these tasks have largely been accomplished. Let us see how it was done.

The statistical-mechanical model of the B-Z transition in DNA with a random base sequence, which was described in Chapter 1, Section III.E, is doubtless the simplest possible one that takes into account the structural features of the Z-form. It contains four energy parameters characterizing the transition of a base pair from the B- to the Z-form, ΔF_{GC}^{I}, ΔF_{AT}^{I}, ΔF_{GC}^{II}, and ΔF_{AT}^{II}. The system of these parameters takes into consideration the dependence of the ΔF value on the type of the pair (AT or CG) and on its conformation in the Z-form. The index I corresponds to a more advantageous state of the pair in energy terms when the urine base is in a *syn* and the pyrimidine base in an *anti* conformation, while the index II stands for the less advantageous state with a *syn* conformation of the pyrimidine and an *anti* conformation of the urine. The two other parameters of the model correspond to the free energy of the boundaries: F_j^{BZ} to the free energy of the junction of the B- and Z-helices, and F_j^{ZZ} to the free energy of the "phase change" in the regular alternation of *syn* and *anti* nucleotide conformations in the Z-helix.

The parameters of this model were determined consecutively on the basis of the analysis of the B-Z transition experiment data in inserts with different sequences. The first step was to determine the transition parameters for inserts with regular sequences. Peck and Wang[218] used d(CG)$_n$ · d(CG)$_n$ (n = 8, 12, 16, and 21) inserts for this purpose, whose transitional energy is governed by two parameters, ΔF_{GC}^{I} and F_j^{BZ}. Indeed, in a fragment with this sequence one can observe such a regular conformation of the Z-helix where all base pairs are in the state I. The transition was registered by way of two-dimensional gel electrophoresis. In all the four plasmids, the transition in the inserts corresponded to the $n \leq m_0$ case, which means that the cooperative transition affected the whole of the fragments. In the follow-up analysis of the experimental data, the authors calculated the transition curves on the basis of statistical-mechanical analysis (see Section VIII), selecting the ΔF_{GC}^{I} and F_j^{BZ} parameters for the best possible agreement between theory and experiment (Figure 65). They came to the conclusion that the observed dependences of δTw (decrease in the equilibrium twist of DNA as a result of the B-Z transition) on σ could not be described within the framework of the general model formulated in Section II, and slightly modified it by introducing an additional parameter. This parameter, *b*, characterizes the unwinding of the double helix at the junctions of the B- and Z-forms. With the inclusion of the parameter, the formula for the free energy of supercoiling in Equation (4.4) acquires the following shape:

$$\Delta G = AN\left[\left(\sigma + \left(m\kappa + 2\gamma_B b\right)/N\right)^2 - \sigma^2\right] \qquad (4.16)$$

The κ parameter in this formula which characterizes the change in the twist of the double helix at the B-Z transition is equal to 1.9. Here are the optimal values of the parameters based on the basis of the experimental data:

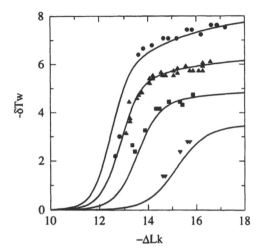

FIGURE 65. The changing equilibrium twist of DNA in the B-Z transition for the $d(CG)_n \cdot d(CG)_n$ insert vs. the linking number difference. Points correspond to experimental data for n equal to 8 (▼), 12 (■), 16 (▲), and 21 (●) curves to theoretical calculations for optimum paramenters. (From Peck, L. J. and Wang, J. C., *Proc. Natl. Acad. Sci. U.S.A.*, 80, 6206, 1983. With permission.)

$\Delta F_{GC}^I = 0.33$ kcal/mol, $F_j^{BZ} = 5$ kcal/mol, and $b = 0.4$. This value of b corresponds to the unwinding of the double helix by 0.4 turns at each B-Z junction. Such a big value of b looks rather strange, considering that the agitation at the junctions of the B and Z helices extends only to 1 or 2 bp.[201] It appears that the b parameter should not be read at face value. Assuming that $b = 0$, the best agreement with the experimental data[218] will be for the same value of ΔF_{GC}^I and a 2-kcal/mol-lesser value of F_j^{BZ}.

The derived value of F_j^{BZ} makes it possible to assess the value of m_0 on the basis of Equation (4.8):

$$m_0 \cong \sqrt{N/3} \tag{4.17}$$

It follows from the comparison of Equations (4.15) and (4.17) that the condition $n \geq m_0$ for the B-Z transition must be met considerably more frequently than in the case of the cruciform structures. Experimentally, this situation was first observed for the transition in the $d(AC)_{30} \cdot d(GT)_{30}$ insert in a circular DNA consisting of 2200 bp.[217] Analysis of these experimental data obtained by way of two-dimensional gel electrophoresis is quite interesting in itself because it effectively illustrates the general treatment for $n \geq m_0$, as outlined in Section II. In this case, only a part of the insert (approximately 25 bp) undergoes a cooperative transition to the Z-form at $\sigma \cong -0.05$. With further growth in supercoiling the length of the Z-helix gradually begins to

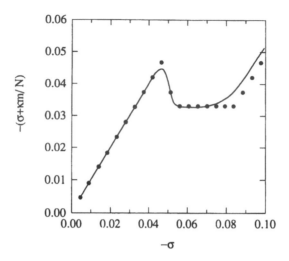

FIGURE 66.   Two-dimensional electrophoresis of DNA with the $d(AC)_{30} \cdot d(GT)_{30}$ insert. The value plotted along the ordinate axis defines the mobility of individual topoisomers. Experimental data from[217] (●) and theoretical calculations for the optimum parameters of the B-Z transition[218] (—) are shown.

expand. Experimental data[217] together with the theoretical curve based on statistical-mechanical calculations[226] are shown in Figure 66. The parameters of the calculation of ΔF, which are, in the terms of this model, $(\Delta F_{GC}^{1} + \Delta F_{AT}^{1})/2$ and F_{j}^{BZ}, were selected on the basis of the optimum conformity of the experimental and theoretical transition curves. The values of the parameters obtained in this manner equalled 0.7 kcal/mol for ΔF and 4 kcal/mol for F_{j}^{BZ}. Knowing the value of ΔF_{GC}^{1}, it is possible to calculate the value of ΔF_{AT}^{1} which proves to equal 1.1 kcal/mol. One should note, however, a slight difference in the ionic conditions used in the studies,[217,218] which may influence the free energy of the B-Z transition. At the same time, the difference in the attained values of the parameter F_{j}^{BZ} should not be given too much attention; it can be explained well by a certain difference in the models of the B-Z transition used and by the inaccurate determination of this parameter.

These data have shown that the $d(CG)_n \cdot d(CG)_n$ sequences assume the left-handed form considerably more easily than the others, which makes them particularly suitable for studying the transition. The following approach was used for finding the other parameters corresponding to various structural elements complicating the formation of Z-DNA. The plasmids had inserts built into them based on the $d(CG)_n$ sequence and containing certain elements upsetting the regular sequence. As a result, the B-Z transition in these inserts happened earlier than other conformational changes and was easy to register. Naturally enough, the number of particular elements included in such inserts

TABLE 1
Thermodynamic Parameters of B-Z Transition

Parameter	Magnitude (kcal/mol)	Ref.
ΔF_{GC}^{I}	0.33	218
ΔF_{AT}^{I}	1.15	228
ΔF_{GC}^{II}	2.6	227
ΔF_{AT}^{II}	3.6	227
F_{j}^{BZ}	5.2	218
F_{j}^{ZZ}	4	228

had to be inversely proportional to the extent to which these elements impeded the B-Z transition. To find the ΔF_{GC}^{II} and ΔF_{AT}^{II} parameters, it is necessary to use sequences with three purines in a row. For example, the free energy of the B-Z transition with the $d[(CG)_3G_2(CG)_3G_2(CG)_3]$ sequence equals $20\Delta F_{GC}^{I} + 2\Delta F_{GC}^{II} + 2F_{j}^{BZ}$. This energy corresponds to the state of a helix with regular alternation of the *syn* and *anti* conformations in each of the helix's chains. The insert may also occur in other states, of course, e.g., when the three $d(CG)_3$ blocks are in the Z-form, while the two dGG nucleotides retain the B-conformation. The free energy of this state will be $18\Delta F_{GC}^{I} + 6F_{j}^{BZ}$, which is most probably higher than the free energy of a Z-helix encompassing the whole insert. The same applies to all other possible states of this insert. In this manner, knowing the ΔF_{GC}^{I} and F_{j}^{BZ} parameters, we can calculate the ΔF_{GC}^{II} parameter from the curve of the transition of this insert. The ΔF_{AT}^{II} parameter is calculated in the same way. The F_{j}^{ZZ} parameter can be determined from the curve of the B-Z transition with a sequence of the $d[(CG)_n(GC)_n]$ type.

The above approach to finding of the parameters of the B-Z transition in irregular sequences was first used by Ellison et al.[227] and completed in the study by Mirkin et al.,[228] in which the full set of six parameters featured for the first time. This set is presented in Table 1.

As suggested by the first experiments involving the B-Z transition, the GC pair in state I has the least free energy in the Z-form. The energy of the AT pair in the Z-form in state I is considerably higher. This explains, among other things, the fact that $d(AT)_n$ sequences transform under the impact of negative supercoiling into cruciform structures rather than into the Z-form (see Section IX). The energy of the AT and GC pairs in state II proves approximately 2 kcal/mol higher than the corresponding values in state I. This means that a certain share of pairs in an inappropriate orientation may occur in zones with a higher-than-average tendency towards the B-Z transition. The energy of the phase change F_{j}^{ZZ} has proved almost the same as the energy of the B-Z boundary, F_j^{BZ}. The comparatively large F_j^{ZZ} value makes frequent phase changes in the Z-form ineffective. The cited parameters correspond to quite definite ionic conditions (TBE-buffer) which were used for the first time in the study by Peck and Wang.[218] All experiments on the basis of which Table 1 has been drawn up were staged under these particular conditions. The expression for the free energy of supercoiling matching this set of parameters is defined by Equation (4.16).

It was very important to make sure that the ΔF values really do not depend on the adjacent neighbors, but are defined mostly by the type and conformation of the given pair. This was checked and confirmed in the study by Mirkin et al. [228] For this, the ΔF_{AT}^l parameter was independently calculated for three types of purine-pyrimidine inserts. In the first insert, the AT pairs contacted only the GC pairs, while in the other two the AT pairs also contacted one another. The values of the ΔF_{AT}^l parameter, found on the basis of independent analysis of the three experiments, coincided and added up to 1.15 kcal/mol, thereby confirming the validity of the assumption. Furthermore, practically the same value of this parameter (1.2 kcal/mol) was obtained in the study[229] which explored the B-Z transition in the d[C(GC)$_6$(AT)$_4$(GC)$_6$G] insert. In that insert, the AT pairs contact almost exclusively one another. In this manner, there are good grounds now for assuming that the above model adequately describes the energy processes attendant to the B-Z transition in DNA with a random sequence.

Two more studies[230,231] deal with the B-Z transition in irregular DNA sequences. Their results do not contradict the picture outlined above, but appear somewhat less coherent.

The presented model of the B-Z transition effectively describes the available experimental data not only for specially constructed circular DNAs, but also for natural sequences. Figure 67 shows a histogram of the probability of occurrence of the Z-form in various parts of ϕX174 DNA (or, more precisely, the probability of anti-Z antibodies binding to a given site on DNA)[193] and the same characteristic calculated theoretically.[232] The calculation corresponding to the ϕX174 DNA sequence was made for approximately the same superhelix density that was used in the experiment. In our opinion, this comparison attests to a good agreement between theory and experiment.

Both the theoretical calculations[232] and the experimental data show that at a superhelix density exceeding 0.06 to 0.07 in absolute terms, Z-regions start to arise in most natural sequences. The number of such regions increases with the growth of $|\sigma|$. As a result, actual torsional stress at $|\sigma| \geq 0.07$ increases very slowly even with the growth of $|\sigma|$. An elegant experimental proof of this behavior of DNA was obtained by Brahms et al.,[56] who carried out a precision study of the change of a DNA's circular dichroism spectra induced by supercoiling. It is clear from Figure 68, which shows the data for two different

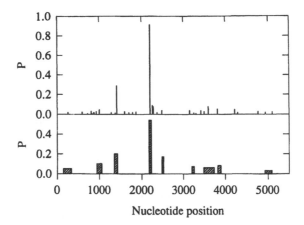

FIGURE 67. Probability of the Z-form occurring in φX174 DNA at σ = −0.06 (top). We cite date[192] which mapped the binding of anti-Z antibodies to this DNA (bottom). Plotted along the ordinate axis here is the probability of antibody binding.

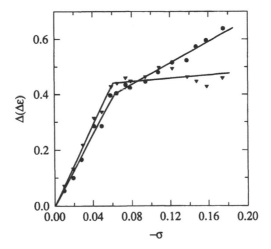

FIGURE 68. Supercoiling-induced changes in the circular dichroism spectra of DNA. The figure shows the dependence of the change Δε on superhelix density for wavelengths of 262 (▼) and 278 nm (●). The bend in the curves marks the beginning of the formation of Z-segments.[56]

wavelengths, that this dependence undergoes a major bend at σ ≅ −0.07. Analysis of the differential spectra of circular dichroism[56] shows that the spectral changes at |σ| ≥ 0.07 are associated with the evolvement of the Z-form. At lower superhelical stresses, changes in the spectra are associated with small changes in the helical rotation angle of the double helix with increasing torsional stress in DNA.

V. MELTING OF CLOSED CIRCULAR DNA

Although research into the formation of noncanonical structures in supercoiled DNA started only in 1980, today we know considerably more about the laws of their formation than about the melting of closed circular DNA, which has been studied for about 20 years now. There are two reasons for this. For one thing, the melting of closed circular DNA is a more involved process compared to, say, the B-Z transition, because the melted state does not have any specific structure. The unwinding of the double helix associated with the denaturation of a DNA fragment must depend on the sign (positive or negative) and the magnitude of superhelical stress. Accordingly, the state of the strands in an open section will determine their entropy and free energy. This considerably complicates the quantitative analysis of the melting of closed circular DNA. The second reason is that the melting process was studied practically without any of the experimental opportunities which have become available to researchers with the advent of genetic engineering.

The first experiments involving the melting of closed circular DNA were staged in the laboratory of Vinograd, California Institute of Technology, Pasadena.[233] The melting was registered by changes in the solution absorbance and in the buoyant density of DNA. It was shown in those studies that the melting of a negatively supercoiled DNA started at considerably lower and ended at considerably higher temperatures than that of the corresponding linear molecules. Such behavior is not surprising. It is clear that as long as the sign of superhelical stress contributes to the unwinding of the double helix, or as long as the degree of denaturation ϑ is less than $(-\sigma)$, this stress should promote further denaturation. At $\vartheta \geq -\sigma$, the melted sections begin to acquire a residual twist because the torsion in the helical regions is no longer sufficient for the realization of the strands' linking number inherent in the molecule, which means that topological constraints impede further melting of the DNA. It is precisely these topological constraints rather than the distribution of the AT and GC base pairs along the chain that have been found to decide the character of the melting of closed circular DNA. This was clearly illustrated in the study[234] which dealt with the melting of closed circular DNA in tetramethyl-ammonium salts at a certain concentration, whereof the melting points of the AT and GC pairs coincide. Under these conditions, the melting interval of a linear DNA narrows down to several tenths of a degree.[235,236] However, the character of the melting of closed circular DNA practically does not change, and the transition remains very broad, starting at 55°C and ending at 110°C (Figure 69). Qualitatively, this melting pattern is described by the theoretical model proposed by Laiken.[237]

The possibility of the formation of open regions under the impact of negative supercoiling must be taken into consideration in the consecutive analysis of

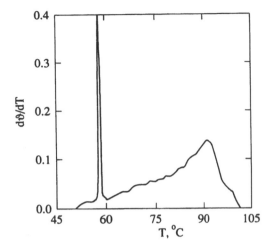

FIGURE 69. The melting of a mixture of closed circular and open forms of φX174 DNA in a tetraethylammonium salt solution, when the melting temperature of the AT and GC pairs is the same. (From Gagua, A. V., Belintsev, B. N., and Lyubchenko, Yu. L., *Nature*, 294, 662, 1981. With permission.) The narrow peak on the differential melting curve corresponds to the melting of the open form, and the wide curve

the emergence of noncanonical structures in DNA, because any upset in the regular structure of the double helix changes superhelical stress. It is essential to know how supercoiling energy depends on the share of open links ϑ, if only at $\vartheta \leq -\sigma$. An attempt to solve this problem was undertaken in the study[238] where the results agree well with the experimental data.[239] This comparison of theory and experiment, however, is very approximate because it is based on the analysis of the initial segments of the complete melting curves without any correlation to the specific sequences of the melting sections.

Unfortunately enough, there are no experimental results yet that would register the melting of well-defined DNA regions identified according to thermal stability. The situation has not changed even after the publication in 1985 of the study by Lee and Bauer, who were the first to apply the method of two-dimensional electrophoresis to the exploration of early melting of closed circular DNA.[240] It was shown in that study that with the rising temperature of the electrophoresis a sequence of mobility leaps appears in the electrophoregram (Figure 70). The authors interpret these leaps as a consecutive cooperative melting of DNA fragments in response to growing negative supercoiling. Although the researchers used a DNA with a known base sequence, it is impossible to identify the exact DNA regions in which the melting takes place because the plasmid in question had no regions of corresponding lengths

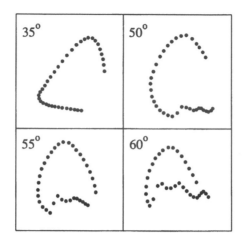

FIGURE 70. Diagram of two-dimensional electrophoresis of circular DNA comprising 5240 bp at elevated temperatures.[240] The observed mobility leaps correspond to the cooperative melting of certain DNA segments.

clearly defined by their thermal stability. Further progress in the study of the melting of circular DNA calls for the use of plasmids with inserts that sharply contrast in terms of their AT content with the sequence in the rest of the circular DNA.

VI. THE H-FORM

It was found in the early 1980s that homopurine-homopyrimidine sequences in negatively supercoiled DNA are hypersensitive to endonuclease S1 which specifically breaks down single-stranded DNA.[162-164] This led to the supposition that a transition to some noncanonical structure, causing a decrease in torsional stress in the molecule, happened in such DNA regions under the influence of supercoiling. However, information on the S1 sensitivity, just as the data on chemical modification,[172] did not allow any definite conclusions to be drawn about this structure at that time, partly because of the divergence in the results of different researchers. Considerable progress in the investigation of this question was made in the studies by Frank-Kamenetskii and co-workers who used the method of two-dimensional electrophoresis to study structural changes in these regions.[241-244]

It was shown, first of all, in the study[241] that a noncanonical structure did actually form in the homopurine-homopyrimidine regions; electrophoregrams of circular DNA registered a mobility leap triggered off by conformational changes in those regions. In the topological sense, the formation of that structure was equivalent to the unwinding of the double helix or to the transition

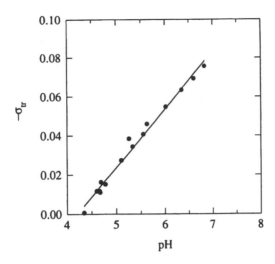

FIGURE 71. Dependence of the superhelix density at which the $d(AG)_{16} \cdot d(CT)_{16}$ adopts the H-form on the solution's pH. (From Lyamichev, V. I., Mirkin, S. M., and Frank-Kamenetskii, M. D., *J. Biomol. Struct. Dyn.*, 3, 667, 1986. With permission.)[242]

of the region into a cruciform structure; the parameter κ was equal to 1. It was ascertained that the superhelix density, σ_{tr}, at which the transition into this structure took place, largely depended on the pH of the solution. The pH dependence of the transition point in the $d(AG)_{16} \cdot d(CT)_{16}$ insert is shown in Figure 71. It follows from this figure that the formation of the structure is stimulated by the rise in the concentration of H^+ ions in the solution. At pH = 4.3, the transition occurs at zero superhelix density. Such an influence of H^+ ions on the transition means that the emerging structure is protonated, which prompted the authors to designate this new structure as the H-form.[241]

In the same study, the researchers carried out theoretical analysis of the formation of a noncanonical structure under the influence of supercoiling with due regard for the protonation of that structure. This analysis resulted in the following formula for the dependence of σ_{tr} on the pH of the solution:

$$\sigma_{tr} = \left(pH_o - pH\right) / \left(10\kappa r\right) \tag{4.18}$$

where r designates the number of base pairs corresponding to a single potential protonation site in the formed structure, and pH_o is a constant. The correlation of Equation (4.18) with the experimental data presented in Figure 71 shows that the parameter r in this case equals 4, which means that there is one protonation site per 4 bp. This conclusion, as well as the derived value of the parameter κ, provided a foundation for developing of the model of the H-form. In this model shown in Figure 72, the homopyrimidine hairpin interacts with

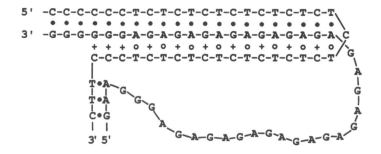

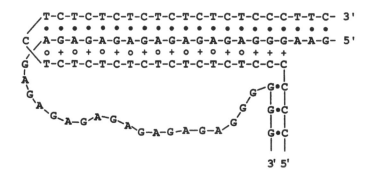

FIGURE 72. Model of the H-form structure proposed by Lyamichev et al.[243] Two isomorphous structures are shown. The pyrimidine strand and half the purine strand make up a triple helix; the other half of the purine strand is free. The CGC triplets are protonated in this structure.

the homopurine chain, forming a triple-stranded complex. The existence of such a triple helix of two antiparallel poly[d(CT)] chains and one poly[d(GA)] chain at low pH was shown earlier[5] (see Chapter 1, Section I.D). In this structure, only the Hoogsteen GC pair is protonated, and overall protonation corresponds to $r = 4$. It is not hard to understand that such a structure may emerge from an open DNA section without any rotation of its ends relative to each other, which means that the parameter κ is equal to 1 regardless of the winding angle of the triple helix.

An elegant confirmation of the accuracy of this model was obtained shortly afterwards.[244] It follows from this model that the H-form can be created by any homopurine-homopyrimidine sequence that is a mirror repeat (H-palindrome). Indeed, in a hairpin formed by the pyrimidine strand, there is always a cytosine opposite a cytosine and a thymine opposite a thymine. To verify this conclusion, four homopurine-homopyrimidine inserts with $d(A_2G_3AGA_2XG_4TATAG_4YA_2GAG_3A_2)$ sequences were synthesized and built

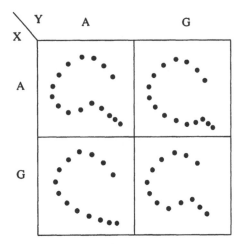

FIGURE 73. Diagram of two-dimensional electrophoresis of four DNAs with d(A$_2$G$_3$AGA$_2$XG$_4$TATAG$_4$YA$_2$GAG$_3$A$_2$) inserts.[244] The values of X and Y are shown in the figure. In DNA molecules with mirror-symmetrical inserts whose electrophoretic patterns are shown in the top left and bottom right corners, the transition to the H-form occurs much earlier than in the other two DNAs (for the case where X = G and Y = A, the transition is not seen at all). All the experiments were performed at pH 4.3.

into plasmids. In the cases when the X = Y = A and X = Y = G, inserts are perfect mirror repeats and must comparatively easily transform into the H-form. In the cases when X = A, Y = G and X = G, Y = A, the symmetry is upset, and the transition to the H-form would be either heavily complicated or blocked altogether. This is precisely the picture that was observed in the experiment (Figure 73). As follows from this picture, in the cases when X = Y = A and X = Y = G, the transition to the H-form is observed at a similar supercoil density. In the case when X = A and Y = G, the transition is heavily shifted into the area of higher superhelical density, while at X = G and Y = A, it does not manifest itself at all. This asymmetry as regards the transition in the latter two plasmids should not be surprising; the insert sequences with X = A, Y = G and X = G, Y = A are not equivalent at all, because the DNA strand is not symmetrical relative to the change of direction. In this sense, the mirror symmetry of the sequence of the H-palindrome is but pseudosymmetry in contrast with the real symmetry of the palindromic area in double-stranded DNA.

A number of further studies involving probing with various chemical agents provided convincing proof of the accuracy of the model presented in Figure 72.[178-180] In particular, in the study,[178] the structure forming in the homopurine-homopyrimidine region was probed with diethylpyrocarbonate which selectively modified unpaired adenines. The modification sites were localized at the level of single nucleotides by way of gel electrophoresis (see Section I.A). It has turned out that only adenines located in one of the halves of the homopurine chain can enter into a reaction with diethylpyrocarbonate. This result fully agrees with the proposed model if out of the two possible isomeric structures presented in Figure 72 only one forms in the end. The equilibrium between the two isomeric forms of H-DNA depends on many circumstances. Htun and Dahlberg seem to have succeeded in registering a wide transition between these forms, which happens with the growth of negative supercoiling.[245]

For a long time, researchers failed to obtain clear results through the localization of S1-sensitive regions. Although nuclease probing was precisely the method that led to the discovery of the unusual structure of the homopurine-homopyrimidine regions, it took some radical improvement of the experiment methods for the results to allow an unequivocal interpretation.[165] Finally, the photofootprinting method[186] (see Section I.A) proved very effective for probing the structure of H-DNA.

Lyamichev et al.[246] determined the energy parameters of the development of the H-form for the $d(CT)_n \cdot d(AG)_n$ and $d(C)_n \cdot d(G)_n$ sequences. As expected, the free energy of the B-H junction proved close to the corresponding value for cruciform structures and came to 18 kcal/mol. The free energy of the B-H transition per base pair for both sequences is in linear dependence on the pH of the solution. However, the evaluation of the free energy of the transition of inserts with $d(A)_n \cdot d(T)_n$ sequences, done on the basis of the results,[246] does not agree with the experimental data. It is much harder to convert this sequence into the H-form than can be assumed from that evaluation. It was only recently that Fox succeeded in this task for an extended $d(A)_{69} \cdot d(T)_{69}$ insert in the presence of magnesium ions.[247] This result shows that in the event of the B-H transition the change in the free energy of a base pair largely depends not only on the type of the given pair, but also on the type of the adjacent ones.

VII. THE INFLUENCE OF SUPERCOILING ON THE B-A TRANSITION

Among the examined conformational changes which can be precipitated by negative supercoiling, the B-A transition holds a special place. The point is that the influence of supercoiling on this transition is very small, although the B-A transition in circular DNA may radically alter the torsional stress in the molecule. We shall give some attention to this question, especially since its various aspects are discussed in literature.[248,249]

The B-A transition stands apart from the other conformational changes examined in this chapter with its very low value of the parameter κ, which means a very small change in the equilibrium twist of the helix per base pair, which accompanies the transition. In solution, the B-A transition in linear DNA

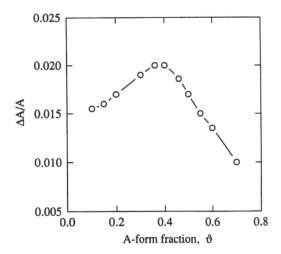

FIGURE 74. Linear dichroism of flow-oriented supercoiled DNA upon A-form fraction.[252] The superhelical density equals –0.012.

is observed at high alcohol concentrations[31,32] and, strictly speaking, the values of γ_B and γ_A (helical repeat of the A- and B-forms) were not known until recently for these conditions. It was natural to assume, however, that these values in alcohol did not differ too much from their corresponding values in solution and in fibers (see Chapter 1, Section I). It followed from those data that $\kappa = 0.05$.

Recently, the κ value was determined directly in an elegant study by Krylov et al.[250] They studied the relaxation of supercoils in superhelical DNA depending on the share of the links which have assumed the A-form. This latter value was determined by changes in the CD spectra, while to register the relaxation of supercoils the researchers measured changes in the linear dichroism of the molecules oriented in the flow. The data obtained in this fashion indicates that the magnitude of linear dichroism passes through its maximum with the rise in the share of links in the A-form increases if the absolute value of superhelix density does not exceed 0.05 (Figure 74). It is clear that the maxima on the curves correspond to the complete relaxation of superhelical stress (with further increase in the degree of the transition the supercoiling changes its sign). The share of links in the A-form, ϑ, corresponding to these maxima, is related to κ by the simple formula:

$$-\kappa\vartheta = \sigma \qquad (4.19)$$

According to the experimental data,[250] $\kappa = 0.044$.

Because of such a small value of κ, the gain in the energy of supercoiling at the B-A transition in circular DNA per base pair proves very small for all reasonable superhelical stress levels. Even if this value, σ, is equal to -0.1, the change in the free energy of supercoiling δG per base pair will be

$$\delta G = AN\left[(\sigma + \kappa / N)^2 - \sigma^2\right] \cong 2A\sigma\kappa \cong 0.1RT \qquad (4.20)$$

Such a small gain in free energy means that supercoiling cannot have any strong influence on the B-A transition. If the transition happens under the impact of the increase in the concentration of alcohol in the concentration range Δa, then the shift in the transition δa as a result of supercoiling meets the condition $\delta a \ll \Delta a$ (see below). At the same time, the B-A transition range itself lies in the area of high alcohol concentrations, and, consequently, supercoiling cannot change in any substantial way the probability of the B-A transition in an ordinary aqueous solution.

A more detailed theoretical analysis of the effect of supercoiling on the B-A transition that occurs with rising alcohol concentration was carried out in a study.[251] The examined model of the B-A transition in closed circular DNA differed radically from the model of the formation of noncanonical structures (described in Section II). This model had to take into consideration the fact that the transition affected the entire molecule and that a lot of regions in the B- and A-forms could arise in it in the transition range. On the other hand, with respect to the B-A transition, DNA can be viewed, in a good approximation, as a homogeneous system. Analysis carried out within the framework of this model led to the following conclusions. The width of the B-A transition in closed circular DNA slightly increases compared to the width of the transition of a linear DNA, while the curve of the transition undergoes a marginal shift (Figure 75). The very form of the transition curve depends on supercoiling, while the concentration of alcohol a_{tr}, corresponding to the middle of the transition, is in linear dependence on the superhelical density:

$$a_{tr} = 2AQ\kappa\sigma / (RT) + constant \qquad (4.21)$$

where Q is a coefficient characterizing the rate of change of the free energy of the B-A transition with changing alcohol concentration,[32] which is equal to $\cong 20$. Assessing the shift of the curve of the B-A transition upon the change of σ in closed circular DNA, one can deduce on the basis of Equation (4.21) that the transition points for two DNAs with different σ values will differ by no more than 1% ($A = 10RT$). It is a value which is hard to register in the experiment. All that can be said is that the experimental data obtained by Krylov et al.[250] do not contradict this assessment; the observed differences in a_{tr} for DNA with different σ values do not exceed 2% by alcohol concentration.

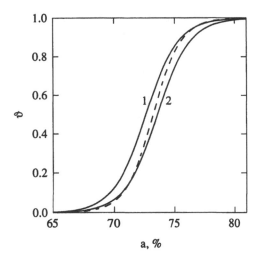

FIGURE 75. Theoretically calculated dependence of the B-A transition in circular (—) and linear (---) DNA upon the concentration of alcohol. Curves 1 and 2 correspond to superhelix densities of –0.05 and 0, respectively.[251]

VIII. STATISTICAL-MECHANICAL ANALYSIS

In the thermodynamic analysis of the formation of noncanonical structures, we took into consideration only two most likely states of the molecule. In the quantitative analysis of conformational changes, however, a lot of possible states with comparable free energy must be considered as a rule. In this section of the book, we shall examine the distinctive features of statistical-mechanical analysis of the formation of noncanonical structures in circular DNA. The general principles of the statistical-mechanical calculation of the characteristics of conformational changes in a DNA molecule were described in brief in Chapter 1, Section III. In the case of circular DNA, the partition function must include all states of the molecule which have a comparable free energy, with due regard for the energy of supercoiling set by Equation (4.5). Sometimes it is easier to do the summation for all states or for a group thereof than to analyze which states should and which should not be taken into consideration. Earlier in the book we examined in sufficient detail the question of the calculation of the free energy of individual states of the molecule. That is why we shall skip the general relations here and concentrate on concrete examples of the statistical-mechanical analysis of conformational transitions in circular DNA, paying special attention to quantitative differences between the results of statistical-mechanical analysis and thermodynamic treatment.

Let us start with the simplest case of the formation of a cruciform structure in the palindromic zone of circular DNA. Let us assume that in the given interval of supercoil density no other noncanonical structures arise in this DNA. In this case, we may confine ourselves to the examination of the various cruciform structures that can arise in this region. Two cases are possible here: (1) the palindromic zone has a marked center (which is typical of most palindromic sequences in native DNA) and (2) in the full palindrome one can distinguish many smaller palindromes with the center displaced in relation to the main palindrome (a sequence of the $d(AT)_n$ type). Let us examine the former case first. The possible states for such a palindrome correspond to the absence of cruciform structures and to crosses of different sizes, with the cruciform structure of each particular size (n, $n - 2$, $n - 4$, ...) realizable in this case only in one way. Upon the formation of a cruciform structure by a region of m base pairs, the change in free energy equals $2F_j + \Delta G(m)$ [see Equation (4.6)], while the statistical weight of the corresponding state, ω_m, is defined by the formula

$$\omega_m = \exp\left[-\left(2F_j + \Delta G(m)\right) / RT\right] \tag{4.22}$$

Assuming that the minimum size of a region out of which a stable cruciform structure can be built is 20 (see Section III), the partition function can be represented as follows:

$$Z = 1 + \sum_m \omega_m \tag{4.23}$$

where the index m decreases via 2 from n to 20. The term 1 in this partition function corresponds to a regular B-conformation of the palindromic zone. The average size of the cruciform structure for a reset superhelix density is defined by the formula

$$\langle m \rangle = \frac{1}{Z} \sum_m m\omega_m \tag{4.24}$$

The Equations (4.23) and (4.24) along with Equation (4.4) provide a comprehensive solution to the problem, with all the calculations easily performed by means of a pocket calculator. Let us assess the relative contribution of different states to the sub-sum in Equation (4.23) or, which is the same thing, the probability of occurrence of these states. If we deal with a typical palindrome of the kind that occurs in native DNA ($n = 30$, $N = 5000$, $\sigma_{tr} = -0.06$), then, as follows from Equation (4.22), the sequence of statistical weights proves close to the geometrical progression with a 0.1 denominator. It is clear that

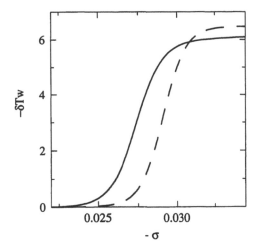

FIGURE 76. Calculated change in equilibrium twist of DNA upon the formation of a cruciform structure in the $d(AT)_{34} \cdot d(AT)_{34}$ insert with all the palindromic states (—) or only two of them (---) taken into consideration.

in this case the partition function can be calculated with the help of the $Z = 1 + \omega_n$ approximation, and the relations supplied by thermodynamic analysis yield an accurate quantitative description of the formation of such cruciform structures. For extended palindromic sequences, though, when the formation of cruciform structures may cause a complete release of superhelical stress ($n \geq m_0$), a whole number of states with similar statistical weights develop. In this case, an accurate description calls for the calculation of the complete partition function after Equation (4.23).

In the event of palindromes of the $d(AT)_n$ type, a cruciform structure consisting of m base pairs can be realized by the $(n - m + 1)$ ways. That is why the partition function in this case acquires the following shape:

$$Z = 1 + \sum_m (n - m + 1)\omega_m \qquad (4.25)$$

where m decreases from n to 20 via 2. Compared to the first case, the number of states increases, and this introduces a considerable adjustment into the equations obtained as a result of thermodynamic analysis. Figure 76 shows the average change in equilibrium twist at the formation of the cruciform structure under review vs. superhelical density. This dependence has been calculated with due regard for all possible states and for only two states corresponding to the largest cross and to the linear form of the palindrome. The calculation was done for $N = 4000$ and the palindrome $d(AT)_{34}$. The δTw value was calculated by the formula

$$\delta Tw = -\frac{1}{Z}\sum_m \frac{m}{\gamma_B}(n - m + 1)\omega_m \qquad (4.26)$$

The difference between the two curves is insignificant, although it does exceed the possible experimental error.

Let us now pass on to the formation of regions in the left-handed Z-form. We shall begin with the case when there is an insert in DNA which passes into this structure earlier than all other regions in the sequence. If the sequence in this insert corresponds to the regular alternation of purines and pyrimidines, then only the 1st state of base pairs in the Z-form can be taken into account. Besides, if the length of the insert does not exceed 100 bp, the possibility of the emergence of more than one region in the Z-form may well be ignored. This means that in this case we must take into consideration only the states with all possible positions of the ends of the Z-form region:

$$Z = 1 + \sum_{j>i}\exp\left[\sum_{k=i}^{j}\Delta F_k^I + 2F_j^{BZ} + \Delta G(j - i + 1)\right] \qquad (4.27)$$

It is naturally implied here that i and j must correspond to pairs in the insert under review. This is precisely the approximation for Z that was used in the study,[218] the only difference being that the length of the region in the Z-form always corresponded to an even number of base pairs. This limitation is quite unessential, even though there is a structural regularity about it; the recurrent unit in the Z-helix is 2 bp, rather than a single base pair as in the B- and A-forms of DNA. The difference in the results of these two models can be compensated with a small adjustment of the parameter $2F_j^{BZ}$.

Including in the partition function terms corresponding to different locations of the Z-region within the area of its likely formation is absolutely essential when $n \geq m_0$. Even for $n \leq m_0$, however, the difference between the results derived from Equation (4.27) and those yielded by the consideration of only the principal terms in the partition function (i.e., corresponding to the thermodynamic analysis) proves appreciable. By way of an example, Figure 77 shows the results of calculations by these two methods of the B-Z transition in a $d(CG)_{16}$ insert built into a 4000-bp-long plasmid. The same approach can be used for analyzing the B-Z transition in inserts with sequences containing a fixed number of point substitutions.[227-229]

In the general case of the B-Z transition in a random DNA sequence, it is hard to identify the dominant groups of states. The solution is to take into consideration in calculating the partition function all possible states of the molecule without a preliminary analysis of their relative statistical weight. The

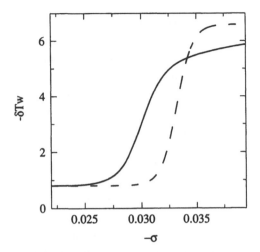

FIGURE 77. Calculated change in equilibrium twist of DNA upon the B-Z transition in the $d(CG)_{16} \cdot d(CG)_{16}$ insert with all possible locations of Z-segment boundaries (—) or only two states of the insert (---) taken into consideration.

overall number of such states, if only the B- and Z-pair conformations are taken into account, is 3^N, and for the chain lengths that are of practical interest it is not possible to calculate the partition function by going through all the possible states. A special effective algorithm making it possible to carry out the appropriate calculations with a computer over a reasonable length of time was developed recently by Anshelevich et al.[232] The number of operations in that algorithm for a molecule comprising N base pairs grows not faster than N^2. The algorithm makes it possible to take into consideration not only the possibility of the formation of fragments in the Z-form, but any other noncanonical structures as well. As will be shown in the next section, in view of the strong mutual influence of the various transitions happening under the impact of negative supercoiling, consistent analysis calls for simultaneous regard for all possible conformational changes which may occur in a given DNA.

IX. MUTUAL INFLUENCE AND COMPETITION OF DIFFERENT STRUCTURAL TRANSITIONS

As repeatedly noted above, there is a strong mutual influence between different transitions in closed circular DNA caused by negative supercoiling. The primary reason for this is that any such transition reduces the superhelical stress in DNA which has caused the transition in the first place, and the shorter the complete length of the DNA, the stronger is this reverse dependence. Indeed, superhelical stress is determined by the sum

$$\left(\sigma + \sum \kappa_i m_i \, / \, N \right) \qquad\qquad (4.28)$$

where the parameters κ_i and m_i characterize the formation of the i-th structure (see Section II). This means that the influence of a structure with a specific torsion and length on superhelical stress is all the smaller, the greater the length of the DNA. In particular, for a DNA consisting of 4000 bp, the formation of a cruciform structure in a palindrome of 30 bp reduces the absolute value in Equation (4.28) almost by 0.01.

Disregard for this mutual influence in the quantitative analysis of conformational changes can substantially distort the results. Let us examine the determination of the energy parameters of the B-Z transition for the $d(CG)_n$ sequence. The two parameters characterizing the B-Z transition in such regions, ΔF_{GC}^{I} and F_{j}^{BZ}, can be determined by studying the corresponding inserts of different lengths built into plasmids. One must be sure, however, that no other structural changes in the molecules under review occur in the superhelix density range where B-Z transitions are observed in these inserts. The most likely additional structural change is the formation of cruciform structures in the palindromic zones of DNA. Under the conditions of the experiments carried out by Nordheim et al.[252] and Singleton et al.[253] the development of a cruciform structure happened precisely in the superhelix density interval in which B-Z transitions were observed (there are direct experimental indications to this effect in Singleton et al.[253]). By ignoring this circumstance, the authors of the above studies ended up with distorted values of the B-Z transition parameters, especially F_{j}^{BZ}.

It is not always easy to monitor such "extra" transitions. In the event of transitions happening at a not too high superhelix density, the absence of "extra" leaps in two-dimensional electrophoregrams can serve as a good control. In the case of transitions corresponding to a high superhelical density, however, this kind of control proves very difficult. One of the possible solutions is to block the formation of cruciform structures by staging experiments at low temperatures when such structures cannot evolve for kinetic reasons.[199] In a number of cases it is more convenient to shift the transition under review, in one way or another, into the area of lower $-\sigma$, e.g., through an appropriate selection of ionic conditions, as was done in the study[218] dealing with the B-Z transition.

The mutual influence of different transitions may lead to some exotic results, such as a nonmonotonic dependence of the probability of the formation of certain structures on superhelix density. In other words, with the rise in negative supercoiling, some of the noncanonical structures which had evolved earlier may revert to the B-form. The possibility of such behavior of circular DNA was first indicated in the study by Benham.[248] The apparent reason for

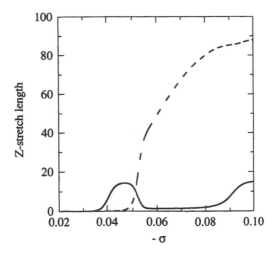

FIGURE 78. Dependence of the average number of base pairs adopting the Z-form in the $d(CG)_8 \cdot d(CG)_8$ (—) and $d(CA)_{60} \cdot d(TG)_{60}$ (---) inserts within a 4400-bp-long circular molecule. Calculations were performed for the values of B-Z transition parameters listed in Table 1.

the reverse transition of a noncanonical structure into the B-form is that another transition takes place in the DNA in the range of stability of this structure, causing a considerable release of superhelical stress. As a result, the existence of the former structure becomes uneconomical. Such behavior, of course, can only occur at specific proportions between the parameters of the two transitions and for a not too great length of the DNA. A detailed theoretical analysis of this question can be found in the study by Vologodskii,[254] and later such a system was investigated experimentally also.[255] The tests involved a specially designed plasmid containing $d(CG)_8 \cdot d(CG)_8$ and $d(CA)_{60} \cdot d(TG)_{60}$ inserts. With the growth of negative supercoiling, the first and then the second, larger, insert assumed the Z-form (Figure 78), which caused a sharp decrease in superhelical stress. This led to a reverse Z-B transition in the first insert. As supercoiling continued to increase, the first insert assumed the Z-form once again. This sequence of transitions cannot be proved with two-dimensional electrophoresis alone, as it yields no information on the location of the transitions. To establish the sequence of transitions, Ellison et al.[255] used a special, very elegant method. A mixture of topoisomers with ΔLk from 0 to –50 was treated with the BssHII restriction endonuclease, which specifically cuts the DNA in the region with the d(GCGCGC) sequence. In the plasmid under review, this sequence could only be found in the first insert. Since the enzyme works only if the corresponding DNA region is in the B-form, only the molecules in which the insert came in the Z-conformation could be retained in the mixture of

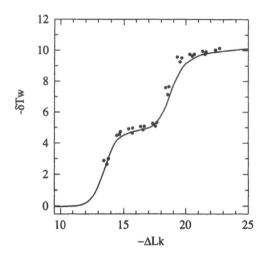

FIGURE 79. Changing equilibrium twist of DNA upon the B-Z transition in two identical inserts, d(CG)$_{12}$ · d(CG)$_{12}$, within the 4400-bp plasmid. (From Kelleher, R. J., III, et al., *Proc. Natl. Acad. Sci. U.S.A.*, 83, 6342, 1986. With permission.) The theoretical curve is shown alongside the experimental data.

topoisomers. Two-dimensional electrophoresis of such a mixture of topoisomers treated with this endonuclease has shown that the d(CG)$_8$ · d(CG)$_8$ insert is in the Z-form in molecules with ΔLk from −17 to −20 and higher than −35. This result perfectly agrees with the theoretical pattern of transitions presented in Figure 78.

Another graphic example of mutual influence is offered by transitions in a plasmid containing two d(CG)$_{12}$ · d(CG)$_{12}$ inserts separated by a certain distance and occurring in the same surroundings.[256] With the growth of negative supercoiling in that DNA, two cooperative transitions were observed, corresponding to conformational changes in both inserts (Figure 79). Since the inserts are identical, it is impossible to say which of them undergoes the transition earlier and which later, but the transition in one of them reduces the superhelical stress and holds back the formation of the Z-helix in the other insert. Such a picture of conformational transitions fully agrees with the calculations carried out within the framework of the B-Z transition model outlined above[256] (see Sections IV and VIII). It must be noted here that this kind of behavior is essentially impossible in linear DNA where two such transitions can happen independently and simultaneously.

Competition between two different structures is possible within one region also. Among the sequences with a regular alternation of purines and pyrimidines which comparatively easily assume the Z-form there may be palindromic sequences which, of course, are capable of forming cruciform structures. Sequences of this type include d(AT)$_n$, d(CG)$_n$, d(CATG)$_n$. The balance between

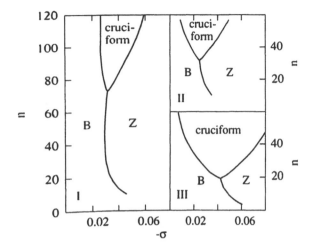

FIGURE 80. Diagrams of the states of inserts with sequences d(CG)$_x$ · d(CG)$_x$ (I), d(CATG)$_x$ · d(CATG)$_x$ (II), and d(AT)$_x$ · d(AT)$_x$ (III) within 4000-bp-long circular DNA molecules. Plotted along the horizontal axis here is the full insert length.

the B-, Z-, and cruciform conformations for such regions depends on their sequence and length and, to a lesser extent, on the length of the whole DNA. This balance can be presented in the form of a diagram of states,[232] as shown in Figure 80. These diagrams were calculated theoretically on the basis of the above models of the formation of noncanonical structures. As can be seen from Figure 80, in the case of short regions only the B- and Z-conformations are stable for all superhelix density values. In the event of sufficient length, however, cruciform structures prove the most stable of all in a certain interval of σ. For regions of such length, one can observe two conformational transitions with the rise in negative supercoiling, the B-form to cruciform and cruciform to the Z-form. The character of these diagrams is associated with the differences in the parameters ΔF, F_j, and κ for the formation of the Z-form and cruciform structures. So far, no one has experimentally observed more than one transition for any of the sequences, although, as follows from Figure 80, this should be possible. Nevertheless, a case of competition between different structures seems to have been observed experimentally in a study of the insert with the d[C(GC)$_6$(AT)$_4$(GC)$_6$G] sequence.[229] The conformation of this insert arising with increasing negative supercoiling manifested the properties of both a cruciform structure and of the Z-helix. According to the theoretical calculations done by the author of this book, the probability of the formation of a cross must be as high as 0.2 in this case, whereupon, with the further rise in supercoiling, it must drop to zero as a result of the ousting of the cruciform structure by the Z-form. This pattern of conformational changes effectively accounts for the experimental data of Ellison et al.[229]

X. KINETICS OF THE FORMATION OF NONCANONICAL STRUCTURES IN CIRCULAR DNA

It was noticed almost immediately after research began into the formation of cruciform structures in circular DNA[187,196,198] that this process is characterized by very long relaxation times. The first quantitative data concerning the kinetics of the formation of cruciform structures was obtained in the course of a study of giant artificial palindromes.[257] According to these data, the period of relaxation in such palindromes came to about 20 min at 25°C. This result, however, did not look so surprising because of the immense size of the emerging structures. Soon it was discovered, however, that the period of relaxation of cruciform structures occurring in a palindrome consisting of 31 bp exceeded the duration of electrophoresis and actually amounted to dozens of hours or more.[198] Surprising in itself, this result contradicted, at first glance anyway, the findings of the study by Mizuuchi et al.,[257] as it appeared that the relaxation time for small structures was considerably longer than the period of relaxation for large ones. Further experiments and theoretical studies have helped to clarify this issue.

By now, the kinetics of the formation of cruciform structures have been studied in sufficient detail and many of its laws, which seemed inexplicable before, have come to be understood. The kinetics of the B-Z transition under the impact of negative supercoiling, however, have not been studied as extensively. There are only two papers dedicated to an experimental analysis of this process.[258,259] These studies have shown that relaxation times, in the case of the B-Z transition, range from minutes to dozens of minutes and quickly decrease with the growth of negative supercoiling (Figure 81). The relaxation processes associated with the development of the H-form have not been specially studied at all. According to the available data,[243] these relaxation times run into hours. That is why we shall deal in detail only with the kinetics of the formation of cruciform structures.

The transition of a palindromic region in DNA from the B-form (B) into a cruciform structure (C) is a first-order process described by the following diagram:

$$B \overset{k_1}{\underset{k_2}{=}} C$$

where k_1 and k_2 stand for the rate constants of the direct and the reverse processes. The establishment of an equilibrium in such a system is described by exponential kinetics with a time constant:

$$\tau = \left(k_1 + k_2\right)^{-1} \tag{4.29}$$

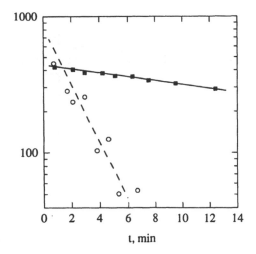

t, min

FIGURE 81. The kinetics of the B-Z transition in the $d(CG)_{16} \cdot d(CG)_{16}$ insert within circular DNA (From Peck, L. J., et al., *J. Mol. Biol.*, 190, 125, 1986. With permission.) We show the time dependence of the fraction of molecules with the insert in the B-form in conventional units. Experimental data were obtained for = −0.051 (○) and for = −0.069 (■).

In this manner, the relaxation time is determined by the maximum rate constant for the given conditions. It must also be noted that the relationship between the k_1 and k_2 constants and the equilibrium constant K is governed by the following formula:

$$K = k_1 / k_2 \qquad (4.30)$$

In the event of the formation of a cruciform structure out of a regular double helix, the process must consist of two stages. First a germ of the new conformation appears, then this germ may develop into a structure of equilibrium dimensions for the given conditions. In the event of a cruciform structure, which is radically different from the B-form, the development of the germ can hardly happen other than through the opening of the corresponding region. Such an open region must correspond to the minimum length at which stable cruciform hairpins, as compared with the open state, may form, which means the inclusion of 10 to 15 bp.[260,261] Fluctuational opening of a region of such length under room temperature in a relaxed DNA is an extremely rare phenomenon, but with the rise in negative supercoiling the probability of such an opening increases very sharply also. According to expert evaluations,[260,261] the formation of the cruciform germ is the limiting phase of the process. The immediate inference is that the rate constant of the formation of cruciform structures, k_1, should not depend on their size, but must strongly depend on

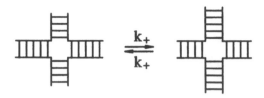

FIGURE 82.   Diagram of the branch point shift in a cruciform structure. Nucleotide sequence permitting, the shift can happen in either direction with equal probability.

the superhelical density. This conclusion is confirmed by the findings of Gellert et al.,[262] which dealt with the kinetics of the formation of cruciform structures in palindromes of different lengths at different superhelix densities. This means that there is no contradiction between the relatively small times of the formation of large crosses[257] and the very large times of the formation of small crosses;[198] all one has to do is assume that the superhelix density of the DNA used by Mizuuchi et al.[257] was higher than that of the DNA used by Lyamichev et al.[198]

The reverse process of the transition of a palindromic region from a cruciform structure to a regular double helix may be very slow also. This process goes after the following pattern. The equilibrium size of the cruciform structure fluctuates down to the size of the germ, which transforms into an open region quickly closing into a B-helix. The decrease in the size of the cruciform structure will occur through a transfer to the main helix of two symmetrical pairs abutting the foot of the cross. This process apparently has a random walk character, which means that it can just as well go in the opposite direction. This kind of random walk of the crossing point of four paired complementary chains was explored from the quantitative angle in the study by Thompson et al.[263] (Figure 82). That study demonstrated that the rate constant of an elementary step k_+ in that process came to about 10^3 sec^{-1}. However, the random wandering of the crossing point in the case of cruciform structures in supercoiled DNA differs from the situation examined by Thompson et al.[263] In this case, the probability of the displacement of the crossing point towards a smaller size of the cross, (p_-), is considerably lower than the probability of the reverse step, (p_+). This is due to the fact that the decrease in the size of the cross corresponds to an increase in the free energy of supercoiling, while the reverse step means a decrease of that energy. As follows from Equation (4.5), a change in the free energy of supercoiling as a result of a single elementary displacement of the crossing point, δG, satisfies the equation

$$\delta G = 40RT\sigma \qquad (4.31)$$

and, consequently,

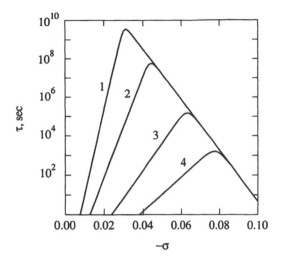

FIGURE 83. The relaxation time of cruciform structures in circular DNA vs. superhelical density.[260] Calculations were performed for a temperature of 25°C and a palindromic GC content close to 0.5. Different curves correspond to different palindrome lengths: 1 corresponds to 65, 2 to 45, 3 to 31, and 4 to 25.

$$p_- \,/\, p_+ = \exp(40\sigma) \tag{4.32}$$

As follows from Equation (4.32), for $\sigma = -0.05$ the ratio p_-/p_+ comes very close to 0.1. This means that a decrease in the size of a cruciform structure by several base pairs will call, on an average, for an immense number of steps and, consequently, for a great amount of time, and this time is bound to increase very sharply with the growth of negative supercoiling. Quantitative analysis of this process[260] has made it possible to determine the rate constant of the disintegration of the cross, k_-, as well as k_+, on the basis of Equation (4.30).

The results of calculations[260] for palindromes of different lengths are presented in Figure 83. The maximum relaxation time for each palindrome corresponds to the point of equilibrium transition between the cruciform and linear forms. At this point, the relaxation time proves enormous for practically all cruciform structures which arise under the impact of negative supercoiling. This time grows sharply as the equilibrium transition point moves to the area of lower superhelix density, i.e., as the palindrome grows in size. In the zone of stability of cruciform structures, however, the time of relaxation (in this case, the time of cross formation) depends solely on the superhelical density. These results show that at room temperature, for which the calculations were performed, the formation of a cruciform structure can only be observed at a sufficiently high superhelical density, exceeding 0.05 in absolute terms. At lower values of $-\sigma$, the transition simply does not have the time to occur within the characteristic duration of an experiment. It should be noted that this conclusion was reached independently by Gellert, O'Dea, and Mizuuchi[262] on the basis of their own experimental data.

The theoretical dependence shown in Figure 83 correlates well with the experimental data from the studies by Panyutin et al.[212] and Courey and Wang,[261] which dealt with the kinetics of cross formation depending on superhelix density.

It has turned out that the time of relaxation is in sharp dependence on the temperature of the solution.[208,212,261,262,264] In particular, at a temperature of several degrees above zero (Celsius), the time of the formation of a cruciform structure even at $\sigma = -0.06$ could amount to weeks and even more.[208,212,262] That is why none of the manipulations with DNA at this temperature shift the proportion between the two forms. This circumstance considerably facilitates the study of the kinetics of cruciform structures both in a solution and in the cell (see Chapter 6).

As shown by Panyutin et al.,[199] the relaxation times in the sequence $d(AT)_n$ have proved incomparably shorter than the times of relaxation of palindromes of the corresponding length with a GC content close to 0.5. This fact correlates with the above qualitative description, because the fluctuational opening of the regions consisting solely of AT pairs happens incomparably more frequently. The question of the influence of the GC composition of the palindrome's center on the kinetics of cross formation was the specific subject of the study by Courey and Wang.[261]

It was found that the speed of an elementary act in the process of cross size reduction was decelerated by several orders of magnitude upon the addition of Mg^{2+} ions to the solution.[264] The nature of this effect is not quite clear. Most probably, the Mg^{2+} ions form very stable complexes at the intersection site of the four helices.[264] In some cases, the kinetics of the formation of cruciform structures depend not only on the characteristics of the palindromic region, but also on its surroundings.[265] It should be noted here that quantitative analysis of the results of some of the experimental studies is difficult because they did not monitor the superhelix density in the DNA preparation. As has been shown above, this parameter is the most important in this case.

Chapter 5

THE USE OF CIRCULAR DNA IN STUDIES OF THE DOUBLE HELIX AND ITS GENERAL PROPERTIES

The special properties of circular DNA not only serve as an object of study, but in many cases provide a unique tool for the study of the double helix itself. We have already encountered some such cases in the previous chapters of this book. For example, Chapter 3, Section II.A dealt with determining the torsional rigidity of the double helix through an analysis of the equilibrium distribution of molecules over topoisomers. In Chapter 3, Section III, we discussed experiments that proved the helical structure of double-stranded DNA. Circular DNA makes it possible to measure very small changes in the helical rotation angle of the double helix, which occur with changing ambient conditions (Chapter 2, Section II.B). In this chapter, we shall briefly consider two more cases in which circular DNA molecules are used for studying the general properties of the double helix.

I. FINDING THE HELICAL REPEAT OF THE DOUBLE HELIX

The first time the helical repeat of the DNA double helix in solution was accurately determined was in Wang's very elegant study.[266] The approach used in that work was based on the following considerations. Consider the equilibrium distribution of circular DNA molecules according to the linking number of strands, Lk (see Chapter 2, Section III). This distribution is obtained upon the ligation of single-stranded breaks in DNA. The distribution maximum is defined by the value of Lk_0, which is equal to N/γ, where N is the number of base pairs in DNA, and γ is the DNA helical repeat in the absence of torsional stress. Unfortunately, there is no way of measuring the value of Lk_0 directly, as one cannot measure the absolute value of Lk for each topoisomer upon electrophoretic separation. However, it is not difficult to find the fractional part of Lk_0, δ, from the electrophoretic pattern (see Figure 24). The most precise method for determining δ is described by Depew and Wang.[50] The value of δ per se does not enable one to determine the repeat of the double helix. Therefore, Wang made use of the fact that with changing DNA length the fractional part δ will undergo oscillations with a period of γ. Thus, what one needs to do is obtain DNA molecules with a reset selection of lengths and then find the equilibrium topoisomeric distribution for them. By the early 1980s, a lot of ground had

TABLE 2
The Helical Repeat of the Various Nucleotide Sequences

Sequence	Helical repeat, γ	References
poly(dA) · poly(dT)	10–10.1	267–269
poly(dAT) · poly(dAT)	10.7–10.8	268, 269
poly(dG) · poly(dC)	10.7	267
Quasirandom	10.5–10.6	61, 267–269

been covered in manipulations with plasmid DNA sequences, and, having performed the appropriate experiments, Wang came up with the first assessment the double helix repeat for DNA in solution (his experimental scheme differed somewhat from the one described above, but was based on the same ideas). This assessment was later refined, and the value of γ averaged over the DNA sequence is currently believed to be 10.55 ± 0.03.[61,267-269] Note that X-ray diffraction data for DNA fibers yield a γ value of 10.0.[1] The discrepancy is most probably due to the packing effects in DNA fibers.

What this approach allows one to establish is not the sequence-average value of γ, but the periodicity in a specific sequence that gradually integrates itself into circular DNA. Therefore, one can examine the dependence of γ on the nucleotide sequence in the double helix. A recent study[269] suggested a way to more accurately measure the fractional part of Lk_o by using specially chosen conditions of electrophoresis, where the mobility of topoisomers is in a strictly linear dependence on the absolute value of the linking number difference in closed circular DNA. This experimental approach to determining the DNA helical repeat for different sequences was used in the works of Peck and Wang,[267] Strauss et al.,[268] and Goulet et al.[269] The results of those studies are summed up in Table 2.

II. FINDING THE DOUBLE HELIX UNWINDING ANGLE UPON LIGAND BINDING

In Chapter 2, Section II.A, we considered the use of ligand binding to circular DNA as a method for determining superhelical density. Elementary analysis shows that superhelical stress would be zero if the number of bound ligand molecules per base pair, ν, satisfies the equation

$$\nu = 360 \cdot \Delta Lk / \phi N \qquad (5.1)$$

where ϕ denotes the double helix unwinding angle upon the binding of one ligand molecule. It is for this ν value that the molecules will have the minimum mobility. If one has an independent way of finding ν, then Equation (5.1) can be used for determining the angle ϕ, since the values of ΔLk and N can be found by other methods. For practical purposes, it is more convenient to measure the shift of the maximum for the equilibrium topoisomeric distribution of DNA molecules, Δ, obtained with the help of topoisomerase I through ligand binding. This approach makes for a considerably more accurate evaluation of the angle ϕ. One will easily see that in this case the value of ϕ is defined by the equation

$$\phi = 360\Delta \,/\, \nu N \qquad\qquad (5.2)$$

where ν stands for the number of bound ligand molecules that has caused the maximum shift equal to Δ. Equations analogous to Equation (5.2) are used for establishing the dependence of the double helix helical rotation angle on the concentration of various ions in solution and on temperature (see Chapter 2, Section II.B). The chief difficulty that one has to deal with when using this approach to find ϕ consists in determining the value of ν. For intercalating ligands with aromatic cycles, one can evaluate ν through spectrophotometric techniques — by using the changed spectral properties of the ligands on binding to DNA.

This approached was applied by Wang et al.[270] to the structural study of a complex between RNA polymerase and DNA. In that case, the number of DNA-bound enzyme molecules was determined through sedimentation. Wang et al.[270] demonstrated the local unwinding of the double helix to come to 240° upon the binding of one RNA polymerase molecule at a moderate ionic strength.

Chapter 6

DNA SUPERCOILING INSIDE THE CELL

I. SUPERHELICAL STRESS

Although DNA molecules isolated from cells have a negative superhelicity sufficient for the formation of many noncanonical structures, the torsional stress of DNA within a cell is not such a simple issue. As it turns out, in the cell DNA can be firmly enough bound to various protein structures which fix the conformation of some segments in the double helix and thus lend a certain writhing and twist value to the molecule. Among such structures, one should primarily cite nucleosomes, which are present in all eukaryotic cells. The core of a nucleosome comprises eight histone proteins with a DNA segment, approximately 175 bp long, tightly coiled around it.[271] DNA makes about two turns in a nucleosome, i.e., each nucleosome adds a value close to –2 to the writhing of a DNA molecule. Even though writhing is not an additive value, for a conformation comprising a set of local structures, relative to the overall size of the molecule, additivity does exist in the first approximation.[272] Therefore, the writhing value defined by the presence of nucleosomes is roughly equal to double the number of nucleosomes, with the sign reversed. In the case of prokaryotes, DNA forms no such firm complexes with protein structures, but it is not in a free state in prokaryotic cells either. The superhelical stress in DNA bound to protein structures is defined by the difference between its linking number difference and the writhing and twist change due to the protein structures to which it is bound. This stress may be considerably less than the stress in the same DNA free of proteins. It cannot be ruled out that superhelical stress changes in the course of a cell's life cycle, but this is not entirely clear yet. Finally, superhelical stress can be dynamic and have an uneven distribution along the DNA molecule, depending on the location of coding sites. This is further discussed in Section III.

In view of the above, it is extremely important to be able to measure the torsional stress of DNA within a cell directly. The problem was first approached by Sinden et al.[273] Their approach was based on measuring the amount of DNA-bound trimethylpsoralen molecules for a specified concentration of the latter in the medium. Upon binding, this ligand intercalates into the double helix and thus unwinds it. Therefore, the binding of trimethylpsoralen to negatively supercoiled DNA is more effective than to relaxed DNA. This enables one, so long as one has a graduation curve, to assess the torsional stress of the DNA in question by the number of bound molecules. The possibility of using this method to estimate torsional stress within a cell is based on two distinctive traits of this ligand. First, it can penetrate into the cell through the membrane. Second, when exposed to 300 nm light, DNA-bound trimethylpsoralen forms photo cross-links, so the amount of bound molecules can be determined after the DNA is isolated from the cell. Of course, the ligand's concentration in the cells must differ from its concentration in the medium, and measuring the concentration inside the cell is very difficult. Sinden et al.[273] dealt with this difficulty by comparing the binding of trimethylpsoralen in intact cells and in cells preexposed to γ-irradiation, which causes single-stranded breaks in DNA. Thus, the ligand's binding to native DNA was experimentally compared to the case of a DNA known to lack torsional stress.

This approach yielded the following results.[273] Superhelical stress in *E. coli* cells corresponds to a superhelix density of 0.05 ± 0.01, while it is totally absent in eukaryotic cells.

Another approach to measuring torsional stress within a cell was recently proposed by Bliska and Cozzarelli.[274] It is based on the topological analysis of the products of site-specific recombination of the λ phage Int system in a specially constructed plasmid. As a result of recombination, this plasmid gives rise to two double-stranded linked circular DNA molecules, whose linking number is in linear dependence on the original plasmid's superhelix density. If one uses this system for estimating torsional stress within a cell, however, one runs into a number of difficulties, one of which is the presence of topoisomerases of type II, which are responsible for the reaction of decatenation (see Chapter 2, Section IV). Essentially, this method supplies only the lower boundary estimate for torsional stress in the cell. The torsional stress value thus obtained corresponds to a superhelix density of –0.025.

Finally, one more approach to this problem, which has been gaining recognition in recent years, is based on the analysis of noncanonical structures forming under the influence of superhelical stress. This matter is worthy of a separate discussion.

II. THE EXISTENCE OF NONCANONICAL STRUCTURES WITHIN A CELL

Many circular DNA molecules isolated from cells reveal cruciform structures which have arisen in the palindromic regions under the influence of negative supercoiling. At the superhelix density characteristic of DNA isolated from cells, a number of sequences adopt the left-handed Z-form. This, however, does not mean that torsional stress in DNA within a cell is sufficient for these noncanonical structures to arise. How does one check whether or not these structures form *in vivo*? It proved to be easier to do this for cruciform structures.

The approach used by Courey and Wang[187] and Lyamichev et al.[275] was based on the very slow kinetics of the formation and dissolution of cruciform structures at temperatures around 0°C. Under these conditions, the relaxation times run into weeks or longer. Therefore, if DNA is isolated from cells at these temperatures, the conformation of palindromic regions must remain unchanged. Lyamichev et al.[275] specially checked that none of the procedures involved in isolating DNA from cells changed the structure of palindromes in supercoiled DNA if performed at these low temperatures. In this way it was demonstrated[275] that the main palindrome of pAO3 DNA growing in *E. coli* cells did not carry a cruciform structure. After a short heating at about 40°C, the molecules did develop cruciform structures. A similar result was obtained by Courey and Wang,[187] who examined the state of an artificial 68-bp palindrome integrated in a plasmid. Thus, in both cases, there were no cruciform structures within a cell. Sinden et al.[276] concluded that there were no cruciform structures inside a cell, having fixed the in-cell conformation of the palindrome by psoralene inter-strand cross-links and analyzed the structure of the palindromic region in isolated DNA.

In principle, there are two possible reasons why palindromes do not adopt cruciform structures within a cell. The result is not surprising for the pAO3 plasmid DNA, as the transition here occurs at a high superhelix density, for physiological ionic conditions.[275] Most probably, the torsional stress corresponding to this density is not realized in the cell (see Section I). The case of the long palindrome[187] is not so simple. This palindrome adopts the cruciform structure at a fairly low superhelix density, so the absence of the cruciform from cells is most probably due to kinetic causes.[187] Indeed, the time of cruciformation is strongly dependent on superhelix density (Chapter 4, Section X). At a superhelical density below 0.04 in absolute terms, cruciform structures should not occur in circular DNA molecules in the characteristic time of an *E. coli* cell generation at 37°C. Though the relaxation times of cruciform structures depend on the nucleotide sequence, this conclusion holds for palindromes with a GC content close to 50%. Thus, though the superhelical density at which large palindromes should undergo an equilibrium transition to the cruciform structure may be less than the torsional stress within a cell, these structures do not arise in the cell for kinetic reasons.[187] This conclusion is confirmed by the *in vivo* conformation analysis of a giant natural palindrome.[277] This palindrome should adopt the cruciform structure at $\sigma = -0.015$ [see Equation (4.9)]. The torsional stress in the cells most probably corresponds to a higher superhelix density. Still, this palindrome was in linear form in the cell. Hence, the kinetic causes prohibiting cruciformation within a cell may prove paramount in some cases.

These results show that ordinary cruciform structures (with a GC content of about 50%) are not likely to evolve in a cell, because the thermodynamic

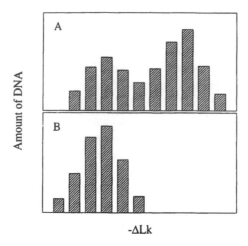

FIGURE 84.   Diagram of topoisomeric distribution of plasmid DNA with the $d(CG)_{12}$ · $d(CG)_{12}$ insert isolated from *E. coli* cells (A) and of the same DNA without the insert (B).[278] The bimodality of the former distribution corresponds to the insert adopting the Z-form in some of the molecules in the cells. Both distributions were obtained after chloramphenicol was added to the cells.

restriction there is compounded by a kinetic one which cannot be overcome with a larger size of the palindrome. In this sense, the situation with the Z-form and cruciform structures occurring in $d(AT)_n$ · $d(AT)_n$ sequences proves more favorable, because their relaxation periods are considerably shorter than those of ordinary cruciforms. It has indeed been shown in a number of studies that under certain conditions such noncanonical structures do form in cells. For analyzing the state of the structures with comparatively short relaxation times, though, a different approach is needed, because their conformational state upon the release of DNA cannot be fixed. What has now become the most wide-spread approach was suggested by Haniford and Pulleyblank.[278] It is based on the analysis of the topoisomeric distribution of a DNA preparation isolated from cells. The point is that the system of cellular topoisomerases maintains superhelical stress at a constant level for a given DNA, which depends on the nucleotide sequence.[53,54,278] That is why in the event of a structural transition in the DNA, causing a drop in torsional stress, the distribution of topoisomers must shift into the area of high degrees of supercoiling, so that torsional stress should revert to its previous level. In experiment, though, instead of the displacement of the whole distribution, its bifurcation was observed more frequently. This kind of bimodal distribution of topoisomers in a plasmid with a $d(CG)_{12}$ · $d(CC)_{12}$ insert and the distribution in a similar reference plasmid without the insert are shown in Figure 84. It has turned out that the distribution of the first group of topoisomers matches the distribution pattern for the reference plasmid, whereas the second group of topoisomers is displaced relative to the first one by a δLk value corresponding to the release of torsional stress resulting from the insert's transition to a left-handed Z-form.[278] This shows that a structural transition has indeed occurred in the part of the molecules with the insert. It must be noted that this result was observed only in cells into which chloramphenicol had been added, suppressing protein synthesis. The suppression of protein synthesis along with certain other factors increases the supercoiling of plasmid DNAs.[213,279] The same method was used for analyzing the state of the cruciform structures in cells, occurring in the $d(AT)_n \cdot d(AT)_n$ inserts.[213,280] These studies showed the existence of cruciform structures in cells subjected to various influences inhibiting their normal growth.

This method of registering the formation of noncanonical structures in a cell can be used for evaluating DNA's torsional stress *in vivo*. Indeed, by finding the minimum length of inserts which assume noncanonical forms in cells under specific conditions, it is possible to evaluate the torsional stress of DNA.[213,281,282] Such analysis, however, must take into consideration the dependence of the superhelix density at which the insert's transition takes place on the ionic conditions. The general assumption is that the ionic conditions in a cell correspond to 0.2 M NaCl, though most experiments to register the formation of noncanonical structures *in vitro* were carried out at a considerably lower concentration of Na^+ ions. This point was taken into account by Dayn et al.[213] who showed that cruciforms in cells treated with chloramphenicol evolved in practically all molecules with the palindrome consisting of 32 and 42 bp and in some of the molecules with a 22-bp palindrome. This means that superhelical stress corresponds under these conditions to such a superhelix density at which the last insert transforms into a cruciform structure under physiological ionic conditions. The latter value measures approximately 0.055.[213] Taking into consideration the rise in superhelix density caused by the addition of chloramphenicol, one can evaluate torsional stress in growing cells at $\sigma = -0.045$.[213] It may be noted here that only Zacharius et al.[282] managed to observe the B-Z transition in growing cells. The evaluation of torsional stress in their study did not take into account the dependence of the transition point on the ionic conditions, and therefore has to be corrected. Bearing in mind this dependence[218] and using the available data,[282] one can arrive at the value of torsional stress in growing cells corresponding to $\sigma = -0.04$.

Thus, different evaluations of torsional stress in prokaryotic cells yield values corresponding to superhelix densities from -0.025 to -0.05. It has become clear now that such a wide range may be due not only to the inaccuracy of the specific measurement methods, but also to the nonuniform distribution of torsional stress in cells, which is discussed in the next section.

III. TRANSCRIPTION-INDUCED DNA SUPERCOILING

After the discovery of DNA gyrase, an enzyme capable of creating negative supercoiling of DNA in *E. coli* cells by using the energy of ATP hydrolysis, it seemed clear that this particular enzyme should play a key role in the formation of superhelices. In 1987, however, Lui and Wang came up with a hypothesis[283] which was convincingly proven later in experiments[284,285] and according to which a major role in the process of supercoiling belongs to the transcription of DNA. In the course of this process, RNA polymerase synthesizes on DNA molecules of mRNA, on which protein chains are then synthesized. Since DNA has the structure of a helix, in the course of this process either the RNA polymerase complex must rotate around the axis of the double helix, or the double helix must turn upon itself. The rotation of the RNA polymerase complex runs into considerable complications because of its very large size; RNA polymerase itself is one of the largest enzymes. Then there is the newly synthesized RNA chain, onto which ribosomes settle at once in a number of cases and protein synthesis gets going. While DNA can turn around its axis, even if that axis forms an involved spatial curve (of the kind formed by the string of the car speedometer), this rotation cannot catch up with the movement of the transcription complex.[286] That is why, as it moves along the DNA, the polymerase complex creates an area of positive supercoiling in front of itself and of negative supercoiling behind itself. Such supercoiling may dissolve as a result of the diffusion of supercoils and their disappearance at the ends of the molecule or as a result of mutual neutralization in the case of circular DNA (Figure 85A). However, for various reasons, such mutual neutralization does not come easy. It happens if the synthesis of RNA occurs simultaneously in two regions of circular DNA in reciprocal directions (Figure 85B). In this case, two regions arise in DNA, one with positive and the other with negative supercoiling, which are separated by RNA polymerase complexes. Another possible reason restricting the diffusion of supercoils is the attachment of DNA at some points to large cellular structures.[287] In such cases, the transcription can continue only if the torsional stress arising with the movement of RNA polymerase is released by the effect of topoisomerases. In *E. coli* cells, two different topoisomerases cause the relaxation of positive and negative supercoiling, while upon the suppression of the activity of one of them, the resultant supercoiling of plasmid DNA isolated from cells must change. Such a change was indeed observed in experiments by Wu et al.[284] and Giaever and Wang.[285] In particular, upon the suppression of the activity of the DNA gyrase responsible for the relaxation of positive supercoils, positive superhelicity was observed in the said plasmid. It was shown in control experiments that upon the suppression of transcription caused by the addition of rifampicin to the

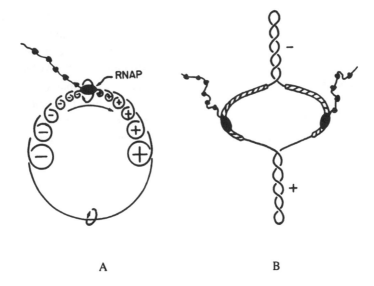

A   B

FIGURE 85. Transcription-induced supercoiling in DNA. In the case of the transcription complex (A) bound to circular DNA, the induced positive and negative supercoils may diffuse towards each other and thus disappear. In the case of two complexes moving towards each other, positively and negatively supercoiled regions arise in the DNA molecule (B). (From Wu, H.-Y., et al., *Cell,* 55, 433, 1988. With permission.)

cells, no such changes in supercoiling were observed. It was also shown[284] that the supercoiling of DNA isolated from cells with a suppressed gyrase activity depended on the number and orientation of promoters in that DNA (promoters are specific regions where transcription begins).

These results show, above all, that topoisomerases are absolutely essential to normal transcription. It is unclear, however, whether transcription in cells with normal topoisomerase activity can lead to the development of a local torsional stress in DNA that is substantially different from the mean value. In cells with normal topoisomerase activity, the supercoiling of isolated DNA changed very insignificantly with changes in transcription intensity and in the number and intensity of promoters.[284] It is possible that topoisomerases very quickly remove the excess torsional stress, so that local torsional stress practically always corresponds to the mean value. It is quite likely, however, that the relaxation of local torsional stress happens at the isolation of DNA, while in the course of transcription it might differ substantially from the mean value. The results of the analysis of the formation of Z-DNA in $d(CG)_n \cdot d(CG)_n$ inserts introduced into plasmids upstream and downstream from the region undergoing transcription, which were obtained by Rahmouni and Wells,[288] clearly point to the latter possibility. In that study, the B-Z transition in cells was registered not by the distribution of topoisomers in the isolated DNA, but by the modification of the B-Z and Z-Z boundaries with osmium tetroxide, which is capable of penetrating into cells (the method of chemical modification in the cell was also used for probing cruciform structures[213,281]). Although the results thus obtained must be adjusted for the ionic conditions (see Section II), they indicate that local torsional stresses may differ by a factor of 2 within the same DNA.[288] It is not very easy to evaluate the correlation between these data and the results of other studies of torsional stress in a cell carried out by similar methods.[213,281,282] One thing is clear — the matter is not closed yet.

REFERENCES

1. **Cantor, C. R. and Schimmel, P. R.,** *Biophysical Chemistry,* Vols. 1–3, Freeman and Co., San Francisco, 1980.
2. **Dickerson, R. E.,** The DNA helix and how it is read, *Sci. Am.,* 249, 94, 1983.
3. **Schlick, T., Hingerty, B. E., Peskin, C. S., Overton, M. L., and Broyde, S.,** Search strategies, minimization algorithms, and molecular dynamics simulations for exploring conformational spaces of nucleic acids, in *Theoretical Biochemistry and Molecular Biophysics, Volume 1: DNA,* Beveridge, D. L. and Lavery, R., Eds., Adenine Press, Schenectady, NY, 1990, 39.
4. **Rich, A., Nordheim, A., and Wang, A. H.-J.,** The chemistry and biology of left-handed Z-DNA, *Annu. Rev. Biochem.,* 53, 791, 1984.
5. **Lee, J. S., Johnson, D. A., and Morgan, A. R.,** Complexes formed by $(pyrimidine)_n \cdot (purine)_n$ DNAs on lowering the pH are three-stranded, *Nucl. Acids Res.,* 6, 3073, 1979.
6. **Kohwi, Y. and Kohwi-Shigematsu T.,** Magnesium ion-dependent triple-helix structure formed by homopurine-homopyrimidine sequences in supercoiled DNA, *Proc. Natl. Acad. Sci. U. S. A.,* 85, 3781, 1988.
7. **Beal, P. A. and Dervan, P. B.,** Second structural motif for recognition of DNA by oligonucleotide-directed triple-helix formation, *Science,* 251, 1360, 1991.
8. **Sundquist, W. I. and Klug, A.,** Telomeric DNA dimerizes by formation of guanine tetrads between hairpin loops, *Nature,* 342, 825, 1989.
9. **Lee, J. S.,** The stability of polypurine tetraplexes in the presence of mono- and divalent cations, *Nucl. Acids Res.,* 18, 6057, 1990.
10. **Koo, H.-S., Wu, H.-M., and Crothers, D. M.,** DNA bending at adenine thymine tracts, *Nature,* 320, 501, 1986.
11. **Volkenshtein, M. V.,** *Configuration Statistics of Polymer Chains,* Interscience, New York, 1963.
12. **Schellman, J. A.,** Flexibility of DNA, *Biopolymers,* 13, 217, 1974.
13. **Hagerman, P. J.,** Flexibility of DNA, *Annu. Rev. Biophys. Chem.,* 17, 265, 1988.
14. **Hagerman, P. J.,** Investigation of the flexibility of DNA using transient electric birefringence, *Biopolymers,* 20, 1503, 1981.
15. **Tanford, C.,** *Physical Chemistry of Macromolecules,* Willy, New York, 1961.
16. **Brian, A. A., Frish, H. L., and Lerman, L. S.,** Thermodynamics and equilibrium sedimentation analysis of the close approach of DNA molecules and a molecular ordering transition, *Biopolymers,* 20, 1305, 1981.

17. **Yarmola, E. G., Zarudnaya, M. I., and Lazurkin, Yu. S.,** Osmotic pressure of DNA solutions and effective diameter of the double helix, *J. Biomol. Struct. Dyn.,* 2, 981, 1985.

18. **Stigter, D.,** Interactions of the highly charged colloidal cylinders with applicatons to double-stranded DNA, *Biopolymers,* 16, 1435, 1977.

19. **Barkley, M. D. and Zimm, B. H.,** Theory of twisting and bending of chain macromolecules; analysis of the fluorescence depolarization of DNA, *J. Chem. Phys.,* 70, 2991, 1979.

20. **Thomas, J. C., Allison, S. A., Appellof, C. J., and Schurr, J. M.,** Torsion dynamics and depolarization of fluorescence of linear macromolecules. II. Fluorescence polarization anisotropy measurements on a clean viral ϕ29 DNA, *Biophys. Chem.,* 12, 177, 1980.

21. **Millar, D. P., Robbins, R. J., and Zewail, A. H.,** Torsion and bending of nucleic acids studied by subnanosecond time-resolved fluorescence depolarization of intercalated dyes, *J. Chem. Phys.,* 76, 2080, 1982.

22. **Robinson, B. H., Lerman, L. S., Beth, A. H., Frisch, H. A., Dalton, L. R., and Auer, C.,** Analysis of double-helix motions with spin-labeled probes: binding geometry and limit of torsional elasticity, *J. Mol. Biol.,* 139, 19, 1980.

23. **Zimm, B. H. and Bragg, J. K.,** Theory of the phase transition between helix and random coil in polypeptide chains, *J. Chem. Phys.,* 31, 526, 1959.

24. **Wada, A., Yabuki, S., and Husimi, Y.,** Fine structure in the thermal denaturation of DNA: high temperature-resolution spectrophotometric studies, *CRC Crit. Rev. Biochem.,* 9, 87, 1980.

25. **Lyubchenko, Y. L., Frank-Kamenetskii, M. D., Vologodskii, A. V., Gauze, G. G., and Lazurkin, Yu. S.,** Fine structure of DNA melting curves, *Biopolymers,* 15, 1019, 1976.

26. **Borovik, A. S., Kalambet, Y. A., Lyubchenko, Y. L., Shitov, V. T., and Golovanov, E. I.,** Equilibrium melting of plasmid ColE1 DNA: electron-microscopic visualization, *Nucl. Acids Res.,* 8, 4165, 1980.

27. **Vedenov, A. A., Dykhne, A. M., and Frank-Kamenetskii, M. D.,** Helix-coil transition in DNA, *Sov. Phys. Usp.,* 14, 715, 1971.

28. **Lazurkin, Yu. S.,** Molecular melting of DNA and the fine structure of melting profiles, *Mol. Biol.,* 11, 1311, 1977.

29. **Gotoh, O.,** Prediction of melting profiles and local helix stability for sequenced DNA, *Adv. Biophys.,* 16, 1, 1983.

30. **Wada, A. and Suyama, A.,** Local stability of DNA and RNA secondary structure and its relation to biological functions, *Prog. Biophys. Mol. Biol.,* 47, 113, 1986.

31. Ivanov, V. I., Minchenkova, L. E., Minyat, E. E., Frank-Kamenetskii, M. D., and Schyolkina, A. K., B to A transition of DNA in solution, *J. Mol. Biol.*, 87, 817, 1974.

32. Ivanov, V. I., Minchenkova, L. E., Minyat, E. E., and Schyolkina, A. K., Cooperative transitions in DNA with no separation of strands, *Cold Spring Harbor Symp. Quant. Biol.*, 47, 243, 1983.

33. Wang, A. H.-J., Quigley, G. J., Kolpak, F. J., Crawford, J. L., van Boom, J. H., van der Marel, G., and Rich, A., Molecular structure of a left-handed double helical DNA fragment at atomic resolution, *Nature*, 282, 680, 1979.

34. Vologodskii, A. V., A theoretical study of the B-Z transition in DNA with arbitrary sequence, *Mol. Biol.*, 19, 876, 1985.

35. Dulbecco, R. and Vogt, M., Evidence for a ring structure of polyoma virus DNA, *Proc. Natl. Acad. Sci. U.S.A.*, 50, 236, 1963.

36. Weil, R. and Vinograd, J., The cyclic helix and cyclic coil forms of polyoma viral DNA, *Proc. Natl. Acad. Sci. U.S.A.*, 50, 730, 1963.

37. Vinograd, J., Lebowitz, J., Radloff, R., Watson, R., and Laipis, P., The twisted circular form of polyoma viral DNA, *Proc. Natl. Acad. Sci. U.S.A.*, 53, 1104, 1965.

38. Coleman, A. and Cook, P. R., Transcription of superhelical DNA from cell nuclei, *Eur. J. Biochem.*, 76, 63, 1977.

39. Benyajati, C. and Worcel, A., Isolation, characterization and structure of the folded interphase genome of *Drosophila melanogaster*, *Cell*, 9, 393, 1976.

40. Ide, T., Nakane, M., Anzai, K., and Andoh, T., Supercoiled DNA folded by non-histone proteins in cultured mammalian cells, *Nature*, 258, 445, 1975.

41. Bauer, W. and Vinograd, J., *Circular DNA. Basic Principles of Nucleic Acids Chemistry*, Vol. 2, Academic Press, New York, 1974, 265.

42. Bauer, W. and Vinograd, J., The interaction of closed circular DNA with intercalative dyes. I. The superhelix density of SV40 DNA in the presence and absence of dye, *J. Mol. Biol.*, 33, 141, 1968.

43. Wang, J. C., The degree of unwinding of the DNA helix by ethidium. I. Titration of twisted PM2 DNA molecules in alkaline cesium chloride density gradients, *J. Mol. Biol.*, 89, 783, 1974.

44. Keller, W., Determination of the number of superhelical turns in simian virus 40 DNA by gel electrophoresis, *Proc. Natl. Acad. Sci. U.S.A.*, 72, 4876, 1975.

45. **Panyutin, I. G., Lyamichev, V. I., and Lyubchenko, Yu. L.,** A sharp structural transition in pAO3 plasmid DNA caused by increased superhelix density, *FEBS Lett.*, 148, 297, 1982.

46. **Lee, C. -H., Mizusawa, H., and Kakefuda, T.,** Unwinding of double-stranded DNA helix by dehydration, *Proc. Natl. Acad. Sci. U.S.A.*, 78, 2838, 1981.

47. **Bauer, W. R.,** Structure and reactions of closed duplex DNA, *Annu. Rev. Biophys. Bioeng.*, 7, 287, 1978.

48. **Wang, J. C.,** Variation of the average rotation angle of the DNA helix and the superhelical turns of covalently closed cyclic λ DNA, *J. Mol. Biol.*, 43, 25, 1969.

49. **Upholt, W. B., Gray, H. B. (Jr.), and Vinograd, J.,** Sedimentation velocity behavior of closed circular SV40 DNA as a function of superhelix density, ionic strength counterion and temperature, *J. Mol. Biol.*, 62, 21, 1971.

50. **Depew, R. E. and Wang, J. C.,** Conformational fluctuations of DNA helix, *Proc. Natl. Acad. Sci. U.S.A.*, 72, 4275, 1975.

51. **Anderson, P. and Bauer, W.,** Supercoiling in closed circular DNA: dependence upon ion type and concentration, *Biochemistry*, 17, 594, 1978.

52. **Shure, M., Pulleyblank, D. E., and Vinograd, J.,** The problems of eukariotic and procariotic DNA packaging and in vivo conformation posed by superhelix density heterogeneity, *Nucl. Acids Res.*, 4, 1183, 1977.

53. **Mirkin, S. M., Zaitsev, E. N., Panyutin, I. G., and Lyamichev, V. I.,** Native supercoiling of DNA: the effects of DNA gyrase and ω protein in *E. coli*, *Mol. Gen. Genet.*, 196, 508, 1984.

54. **Goldstein, E. and Drlica, K.,** Regulation of bacterial DNA supercoiling: plasmid linking numbers vary with growth temperature, *Proc. Natl. Acad. Sci. U.S.A.*, 81, 4046, 1984.

55. **Snounou, G. and Malcolm, A. D. B.,** Production of positively supercoiled DNA by netropsin, *J. Mol. Biol*, 167, 211, 1983.

56. **Brahms, S., Nakasu, S., Kikuchi, A., and Brahms, G.,** Structural changes in positively and negatively supercoiled DNA., *Eur. J. Biochem.*, 184, 297, 1989.

57. **Brown, P. O. and Cozzarelli, N. R.,** A sign inversion mechanism for enzymatic supercoiling of DNA, *Science*, 206, 1081, 1979.

58. **Bauer, W. and Vinograd, J.,** The interaction of closed circular DNA with intercalative dyes. II. The free energy of superhelix formation in SV 40 DNA, *J. Mol. Biol.*, 47, 419, 1970.

59. **Hsieh, T.-S. and Wang, J. C.,** Thermodynamic properties of superhelical DNAs, *Biochemistry* 14, 527, 1975.

60. Pulleyblank, D. E., Shure, M., Tang, D., Vinograd, J., and Vosberg, H.-P., Action of nicking-closing enzyme on supercoiled and nonsupercoiled closed circular DNA: formation of Boltzmann distribution of topological isomers, *Proc. Natl. Acad. Sci. U.S.A.*, 72, 4280, 1975.

61. Horowitz, D. S. and Wang, J. C., The torsional rigidity of DNA and the length dependence of the free energy of DNA supercoiling, *J. Mol. Biol.*, 173, 75, 1984.

62. Shore, D. and Baldwin, R. L., Energetics of DNA twisting. II. Topoisomer analysis, *J. Mol. Biol.*, 170, 983, 1983.

63. Klenin, K. V., Vologodskii, A. V., Anshelevich, V. V., Dykhne, A. M., and Frank-Kamenetskii, M. D., Computer simulation of DNA supercoiling, *J. Mol. Biol.*, 217, 413, 1991.

64. Wang, J. C., Interaction between DNA and an *E. coli* protein ω, *J. Mol. Biol.*, 217, 413, 1971.

65. Gellert, M., Mizuushi, K., O'Dea, M. H., and Nash, H. A., An enzyme that introduces superhelical turns into DNA, *Proc. Natl. Acad. Sci. U.S.A.*, 73, 3872, 1976.

66. Liu, L. F., Liu, C. C., and Alberts, B. M., Type II DNA topoisomerases: enzymes that can unknot a topologically knotted DNA molecule via a reversible double-strand break, *Cell*, 19, 697, 1980.

67. Gellert, M., DNA topoisomerases, *Annu. Rev. Biochem.*, 50, 879, 1981.

68. Liu, L. F., DNA topoisomerases — enzymes that catalise the breaking and rejoining of DNA, *CRC Crit. Rev. Biochem.*, 15, 1, 1983.

69. Wang, J. C., DNA topoisomerases, *Annu. Rev. Biochem.*, 54, 665, 1985.

70. Hsieh, T., Mechanistic aspects of type-II DNA topoisomers, in *DNA Topology and Its Biological Effects*, Cozzarelli, N. R. and Wang, J. C., Eds., Cold Spring Harber Laboratory Press, Cold Spring Harbor, NY, 1990, 243.

71. Baldi, M. I., Benedetti, P., Mattoccia, E., and Tocchini-Valentini, G. P., In vitro catenation and decatenation of DNA and a novel eucaryotic ATP-dependent topoisomerase, *Cell*, 20, 461, 1980.

72. Hsieh, T. S. and Brutlag, D., ATP-dependent DNA topoisomerase from Drosophila melanogaster reversible catenates duplex DNA rings, *Cell*, 21, 115, 1980.

73. Kreuzer, K. N. and Cozzarelli, N. R., Formation and resolution of DNA catenanes by DNA gyrase, *Cell*, 20, 245, 1980.

74. Mizuuchi, K., Fisher, L. M., O'Dea, M. H., and Gellert, M., DNA gyraze action involves the introduction of transient double-strand breaks into DNA, *Proc. Natl. Acad. Sci. U.S.A.*, 77, 1847, 1980.

75. **Borst, P. and Hoeijmakers, J. H. J.,** Kinetoplast DNA, *Plasmid.,* 2, 20, 1979.

76. **Marini, J. C., Miller, K. G., and Englund, P. T.,** Decatenation of kinetoplast DNA by topoisomerases, *J. Biol. Chem.,* 255, 4976, 1980.

77. **Tse, Y. C. and Wang, J. C.,** *E coli* and *M. luteus* DNA topoisomerase I can catalyze catenation or decatenation of double-stranded DNA rings, *Cell,* 22, 269, 1980.

78. **Brown, P. O. and Cozzarelli, N. R.,** Catenation and knotting of duplex DNA by type I topoisomerases: a mechanism parallel with type II topoisomerases, *Proc. Natl. Acad. Sci. U.S.A.,* 78, 843, 1981.

79. **Kikuchi, A. and Asai, K.,** Reverse gyrase — a topoisomerase which introduces positive superhelical turns into DNA, *Nature,* 309, 677, 1984.

80. **Nakasu, S. and Kikuchi, A.,** Reverse gyrase; ATP-dependent type I topoisomerase from Sulfolobus, *EMBO J.,* 4, 2705, 1985.

81. **Forterre, P., Mirambeau, G., Jaxel, C., Nadal, M., and Duguet, M.,** High positive supercoiling in vitro catalyzed by an ATP and polyethylene glycol-stimulated topoisomerase from Sulfolobus acidocaldarius *EMBO J.,* 8, 2123, 1985.

82. **Slesarev, A. I.,** Positive supercoiling catalysed in vitro by ATP-dependent topoisomerase from desulfurococcus amylolyticus, *Eur. J. Biochem.,* 173, 395, 1988.

83. **Nadal, M., Mirambeau, G., Forterre, P., Reiter, W.-D, and Duguet, M.,** Positively supercoiled DNA in a virus-like particle of an archaebacterium, *Nature,* 321, 256, 1986.

84. **Champoux, J. J.,** Mechanistic aspects of type-I topoisomerases, in *DNA Topology and Its Biological Effects,* Cozzarelli, N. R. and Wang J. C., Eds. Cold Spring Harbor Laboratory Press, Cold Spring Harbor, NY, 1990, 217.

85. **Kikuchi, A.,** Reverse gyrase and other archaebacterial topoisomerases, in *DNA Topology and Its Biological Effects,* Cozzarelli, N. R. and Wang, J. C., Eds., Cold Spring Harbor Laboratory Press, Cold Spring Harbor, NY, 1990, 285.

86. **White, J. H.,** Self-linking and the Gauss integral in higher dimensions, *Am. J. Math.,* 91, 693, 1969.

87. **Calugareanu, G.,** Sur las classes d'isotopie des noeuds tridimensionnels et leurs invariants, *Czech. Math. J.,* 11, 588, 1961.

88. **Fuller, F. B.,** The writhing number of a space curve, *Proc. Natl. Acad. Sci. U.S.A.,* 68, 815, 1971.

89. **Crick, F. H. C.,** Linking numbers and nucleosomes, *Proc. Natl. Acad. Sci. U.S.A.,* 73, 2639, 1976.

90. **Edwards, S. F.,** Statistical mechanics with topological constrains: II, *J. Phys. A: Gen. Phys.,* 1, 15, 1968.

91. **Blaschke, W.,** *Introduction to Differential Geometry,* Springer-Verlag, Berlin, 1950.

92. **White, J. H. and Bauer, W. R.,** Calculation of the twist and the writhe for representative models of DNA, *J. Mol. Biol.,* 189, 329, 1986.

93. **Frank-Kamenetskii, M. D. and Vologodskii, A. V.,** Topological aspects of the physics of polymers: the theory and its biophysical applications, *Sov. Phys. Usp.* 24, 679, 1981.

94. **Vologodskii, A. V., Anshelevich, V. V., Lukashin, A. V., and Frank-Kamenetskii, M. D.,** Statistical mechanics of supercoils and the torsional stiffness of the DNA double helix, *Nature,* 280, 294, 1979.

95. **Frank-Kamenetskii, M. D., Lukashin, A. V., Anshelevich, V. V., and Vologodskii, A. V.,** Torsional and bending rigidity of double helix from data on small DNA rings, *J. Biomol. Struct. Dyn.,* 2, 1005, 1985.

96. **Benham, C. J.,** The statistics of superhelicity, *J. Mol. Biol.,* 123, 361, 1978.

97. **Le Bret, M.,** Monte Carlo computation of supercoiling energy, the sedimentation constant, and the radius of gyration of unknotted and knotted circular DNA, *Biopolymers,* 19, 619, 1980.

98. **Chen, Y.,** Monte Carlo study of freely jointed ring polymers. II. The writhing number, *J. Chem. Phys.,* 75, 2447, 1981.

99. **Klenin, K. V., Volgodskii, A. V., Anshelevich, V. V., Dykhne, A. M., and Frank-Kamenetskii, M. D.,** Effect of excluded volume on topological properties of circular DNA, *J. Biomol. Struct. Dyn.,* 5, 1173, 1988.

100. **Shimada, J. and Yamakawa, H.,** Moments for DNA topoisomers: the helical wormlike chain, *Biopolymers,* 27, 657, 1988.

101. **Klenin, K. V., Vologodskii, A. V., Anshelevich, V. V., Klishko, V. Y., Dykhne, A. M., and Frank-Kamenetskii, M. D.,** Variance of writhe for wormlike DNA rings with excluded volume, *J. Biomol. Struct. Dyn.,* 6, 707, 1989.

102. **Benham, C. J.,** Geometry and mechanics of DNA superhelicity, *Biopolymers,* 22, 2477, 1983.

103. **Le Bret, M.,** Twist and writhing in short circular DNAs according to first-order elasticity, *Biopolymers,* 23, 1835, 1984.

104. **Binder, K.,** *Monte Carlo Method in Statistical Physics,* Binder, K., Ed., Springer-Verlag, Berlin, 1979, chap. 1.

105. **Hao, M. H. and Olson, W. K.,** Global equilibrium configurations of supercoiled DNA, *Macromolecules,* 22, 3292, 1989.

106. **Tan, R. K. Z. and Harvey, S. C.,** Molecular mechanics model of supercoiled DNA, *J. Mol. Biol.,* 205, 573, 1989.

107. **Vologodskii, A. V., Levene, S. D., Klenin, K. V., Frank-Kamenetskii, M. D., and Cozzarelli, N. R.,** Conformational and thermodynamic properties of supercoiled DNA, *J. Mol. Biol.,* in press.

108. **Laudon, G. H. and Griffith, J. D.,** Curved helix segments can uniquely orient the topology of supertwisted DNA, *Cell,* 52, 545, 1988.

109. **Boles, T. C., White, J., and Cozzarelli, N. R.,** The structure of plectonemically supercoiled DNA, *J. Mol. Biol.,* 213, 931, 1990.

110. **Adrian, M., ten Heggeler-Bordier, B., Wahli, W., Stasiak, A. Z., Stasiak, A., and Dubochet, J.,** Direct visualisation of supercoiled DNA molecules in solution, *EMBO J.,* 9, 4551, 1990.

111. **Brady, G. W., Satkovski, M., Foos, D., and Benham, C. J.,** Environmental influences on DNA superhelicity, *J. Mol. Biol.,* 195, 185, 1987.

112. **Langowski, J.,** Salt effects on internal motions of superhelical and linear pUC8 DNA. Dynamic light scattering studies, *Biophys. Chem.,* 27, 263, 1987.

113. **Torbet, J. and DiCapua, E.,** Supercoiled DNA is interwound in liquid crystalline solutions, *EMBO J.,* 8, 4351, 1989.

114. **Seidl, A. and Hinz, H.-J.,** The free energy of DNA supercoiling is enthalpy-determined, *Proc. Natl. Acad. Sci. U.S.A.,* 81, 1312, 1984.

115. **Wu, R. and Taylor, E.,** Complete nucleotide sequence of cohesive ends of bacteriophage λ DNA, *J. Mol. Biol.,* 57, 491, 1971.

116. **Wang, J. C. and Davidson, N.,** Thermodynamic and kinetic studies on the interconversion between the linear and circular forms of phage lambda DNA, *J. Mol. Biol.,* 15, 111, 1966.

117. **Wang, J. C. and Davidson, N.,** On the probability of ring closure of lambda DNA, *J. Mol. Biol.,* 19, 469, 1966.

118. **Wang, J. C.,** Cyclization of coliphage 186 DNA, *J. Mol. Biol.,* 28, 403, 1967.

119. **Wang, J. C. and Davidson, N.,** Cyclization of phage DNAs, *Cold Spring Harbor Symp. Quant. Biol.,* 23, 409, 1968.

120. **Jacobson, P. and Stockmayer, W. H.,** Intramolecular reaction in polycondensation. I. Theory of linear systems, *J. Chem. Phys.,* 18, 1600, 1950.

121. **Dugaiczyk, A., Boyer, H. W., and Goodman, H. M.,** Ligation of EcoR1 endonuclease-generated DNA fragments into linear and circular structures, *J. Mol. Biol.,* 96, 171, 1975.

122. **Yamakawa, H. and Stockmayer, W. H.,** Statistical mechanics of wormlike chains. II. Excluded volume effects, *J. Chem. Phys.,* 57, 2843, 1972.

123. **Shore, D., Langowski, J., and Baldwin, R. L.,** DNA flexibility studied by covalent closure of short fragments into circles, *Proc. Natl. Acad. Sci. U.S.A.,* 78, 4833, 1981.

124. **Shore, D. and Baldwin, R. L.,** Energetics of DNA twisting. I. Relation between twist and cyclization probability, *J. Mol. Biol.,* 170, 957, 1983.

125. **Shimada, J. and Jamakawa, H.,** Statistical mechanics of DNA topoisomers. The helical wormlike chain, *J. Mol. Biol.,* 184, 319, 1985.

126. **Levene, S. D. and Crothers, D. M.,** Ring closure probabilities for DNA fragments by Monte Carlo simulation, *J. Mol. Biol.,* 189, 61, 1986.

127. **Hagerman, P. J.,** Analysis of the ring-closure probabilities of isotropic wormlike chains: application to duplex DNA, *Biopolymers,* 24, 1881, 1985.

128. **Hager, P. J. and Ramadevi, V. A.,** Application of the method of phage T4 DNA ligase-catalyzed ring-closure to the study of DNA structure. I. Computational analysis, *J. Mol. Biol.,* 212, 351, 1990.

129. **Taylor, W. H. and Hagerman, P. J.,** Application of the method of phage T4 DNA ligase-catalyzed ring-closure to the study of DNA structure, II. NaCl-dependence of DNA flexibility and helical repeat, *J. Mol. Biol.,* 212, 363, 1990.

130. **Reidemeister, K.,** *Knotentheorie Ergebnisse der Mathematik,* Springer, Berlin, 1932, 74.

131. **Conway, J. H.,** An enumeration of knots and links, and some of their algebraic properties, in *Computational Problems in Abstract Algebra,* Pergamon Press, Oxford, 1970, 329.

132. **Perco, K. A.,** On the classification of knots, *Proc. Am. Math. Soc.,* 45, 262, 1974.

133. **Delbruck, M.,** Knotting problems in biology, in *Mathematical Probelms in the Biological Sciences,* Bellman, R. E., Ed., Am. Math. Soc., Providence, RI, 1962, 55.

134. **Vologodskii, A. V., Lukashin, A. V., Frank-Kamenetskii, M. D., and Anshelevich, V. V.,** Problem of knots in statistical mechanics of polymer chains, *Sov. Phys. JETP,* 39, 1059, 1974.

135. **Frank-Kamenetskii, M. D., Lukashin, A. V., and Vologodskii, A. V.,** Statistical mechanics and topology in polymers chains, *Nature,* 258, 398, 1975.

136. **Michels, J. P. J. and Wiegel, F. W.,** On the topology of a polymer ring, *Proc. R. Soc. London A,* 403, 269, 1986.

137. **Des Cloizeaux, J. and Metha, M. L.,** Topological constraints on polymer rings and critical indices, *J. Phys.,* 40, 665, 1979.

138. **Michels, J. P. J. and Weigel, F. W.,** Probability of knots in polymer ring, *Phys. Lett. A,* 90, 381, 1982.

139. **Koniaris, K. and Muthukumar, M.,** Knottedness in ring polymers, *Phys. Rev. Lett.,* 66, 2211, 1991.

140. **Vologodskii, A. V., Lukashin, A. V., and Frank-Kamenetskii, M. D.,** Topological interaction of polymer chains, *Sov. Phys. JETP,* 40, 932, 1975.

141. **Iwata, K.,** Topological distribution functions of ring polymers II. Theory and computer simulation, *J. Chem. Phys.,* 78, 2778, 1983.

142. **Landau, L. D. and Lifshitz, E. M.,** *Statistical Physics,* Addison-Wesley, Reading, MA, 1958.

143. **Frisch, H. L. and Wasserman, E.,** Chemical topology, *J. Am. Chem. Soc.* 83, 3789, 1961.

144. **Hudson, B. and Vinograd, J.,** Catenated circular DNA molecules in Hela cell mitochondria, *Nature,* 216, 647, 1967.

145. **Clayton, D. A. and Vinograd, J.,** Circular dimer and catenate forms of mitochondrial DNA in human leukaemic leucocytes, *Nature,* 216, 652, 1967.

146. **Krasnow, M. A., Stasiak, A., Spengler, S. J., Dean, F., Koller, T., and Cozzarelli, N. R.,** Determination of the absolute handedness of knots and catenanes of DNA, *Nature,* 304, 559, 1983.

147. **Wasserman, S. A. and Cozzarelli, N. R.,** Biochemical topology: applications to DNA recombination and replication, *Science,* 232, 951, 1986.

148. **Liu, L. F., Depew, R. E., and Wang, J. C.,** Knotted single-stranded DNA rings: a novel topological isomer of circular single-stranded DNA formed by treatment with *E. coli* ω protein, *J. Mol. Biol.,* 106, 439, 1976.

149. **Shishido, K., Komiyama, N., and Ikawa, S.,** Increased production of a knotted form of plasmid pBR322 DNA in *Escherichia coli* DNA topoisomerase mutants, *J. Mol. Biol.,* 195, 215, 1987.

150. **Liu, L. F. and Davis, J. L.,** Novel topologically knotted DNA from bacteriophage P4 capsids: studies with DNA topoisomerases, *Nucl. Acids Res.,* 9, 3979, 1981.

151. **Wasserman, S. A., White, J. H., and Cozzarelli, N. R.,** The helical repeat of double-stranded DNA varies as a funtion of catenation and supercoiling, *Nature,* 334, 448, 1988.

152. **Vologodskii, A. V. and Cozzarelli, N. R.,** A Monte Carlo analysis of the conformation of DNA catenanes, *J. Mol. Biol.,* in press.

153. **Wang, J. C. and Schwartz, H.,** Noncomplementarity in base sequences between the cohesive ends of coliphage 186 and λ and the formation of interlocked rings between the two DNAs, *Biopolymers,* 5, 953, 1967.

154. **Cozzarelli, N. R., Krasnow, M. A., Gerrard, S. P., and White, J. H.,** A topological treatment of recombination and topoisomerases, *Cold Spring Harbor Symp. Quant. Biol.,* 49, 383, 1984.

155. **Wasserman, S. A., Dungan, J. M., and Cozzarelli, N. R.,** Discovery of a predicted DNA knot substantiates a model for site-specific recombination, *Science,* 229, 171, 1985.

156. **Spengler, S. J., Stasiak, A., and Cozzarelli, N. R.,** The stereostructure of knots and catenanes produced by λ integrative recombination: implication for mechanism and DNA structure, *Cell,* 42, 325, 1985.

157. **Kanaar, R., Klippel, A., Shekhtman, E., Dungan, J. M., Kahmann, R., and Cozzarelli, N. R.,** Processive recombination by the phage Mu Gin system: implication for the mechanisms of DNA strand exchange, DNA site alignment, and enhancer action, *Cell,* 62, 353, 1990.

158. **Heichman, K. A., Moskowitz, I. P. G., and Johnson, R. C.,** Configuration of DNA strands and mechanism of strand exchange in the Hin invertasome as revealed by analysis of recombinant knots, *Genes Development,* 5, 1622, 1991.

159. **Singleton, C. K., Klysik, J., Stirdivant, S. M., and Wells, R. D.,** Left-handed Z-DNA is induced by supercoiling in physiological ionic conditions, *Nature,* 299, 312, 1982.

160. **Lilley, D. M. J.,** The inverted repeat as a recognizable structural feature in supercoiled DNA molecules, *Proc. Natl. Acad. Sci. U.S.A.,* 77, 6468, 1980.

161. **Panayotatos, N. and Wells, R. D.,** Cruciform structures in supercoiled DNA, *Nature,* 289, 466, 1981.

162. **Hentschel, C. C.,** Homocopolymer sequences in the spacer of a sea urchin histone gene repeat are sensitive to S1 nuclease, *Nature,* 295, 714, 1982.

163. **Larsen, A. and Weintraub, H.,** An altered DNA conformation detected by S1 nuclease occurs at specific regions in active chick globin chromatin, *Cell,* 29, 609, 1982.

164. **Htun, H., Lund, E., and Dahlberg, J. E.,** Human U1 RNA genes contain an unusually sensitive nuclease S1 cleavage sites within the conserved 3'flanking region, *Proc. Natl. Acad. Sci. U.S.A.,* 81, 7288, 1984.

165. **Belotserkovskii, B. P., Veselkov, A. G., Filipov, S. A., Dobrynin, V. N., Mirkin, S. M., and Frank-Kamenetskii, M. D.,** Formation of intramolecular triplex in homopurine-homopyrimidine mirror repeats with point substitutions, *Nucl. Acids Res.,* 18, 6621, 1990.

166. **Camilloni, G., Seta, F. D., Negri, R., Ficca, A. G., and Di Mauro, E.,** Structure of RNA polymerase II promoters. Conformational alterations and template properties of circularized Saccharomyces cerevisiae GAL1-GAL10 devergent promoters, *EMBO J.,* 5, 763, 1986.

167. **Lilley, D. M. J. and Kemper, B.,** Cruciform-resolvase interactions in supercoiled DNA, *Cell,* 36, 413, 1984.

168. **Johnston, B. H. and Rich, A.,** Chemical probes of DNA conformation: detection of Z-DNA at nucleotide resolution, *Cell,* 42, 713, 1985.

169. **Herr, W.,** Diethyl pyrocarbonate: a chemical probe for secondary structure in negatively supercoiled DNA, *Proc. Natl. Acad. Sci. U.S.A.,* 82, 8009, 1985.

170. **Furlong, J. C. and Lilley, D. M. J.,** Highly selective chemical modification of cruciform loops by diethyl pyrocarbonate, *Nucl. Acids Res.,* 14, 3995, 1986.

171. **Scholten, P. M. and Nordheim, A.,** Diethyl pyrocarbonate: a chemical probe for DNA cruciforms, *Nucl. Acids Res.,* 14, 3981, 1986.

172. **Evans, T. and Efstratiadis, A.,** Sequence-dependent S1 nuclease hypersensitivity of a heteronomous DNA duplex, *J. Biol. Chem.,* 261, 14771, 1986.

173. **Lilley, D. M. J. and Palecek, E.,** The supercoil-stabilised cruciform of ColE1 is hyper-reactive to osmium tetroxide, *EMBO J.,* 3, 1187, 1984.

174. **Galazka, G., Palecek, E., Wells, R. D., and Klysik, J.,** Site-specific OsO_4 modification of the B-Z junctions formed at the $(dA-dC)_{32}$ region in supercoiled DNA, *J. Biol. Chem.,* 261, 7093, 1986.

175. **Lilley, D. M. J.,** Structural perturbation in supercoiled DNA: hypersensitivity to modification by a single-strand-selective chemical reagent conferred by inverted repeat sequences, *Nucl. Acids Res.,* 11, 3097, 1983.

176. **Gough, G. W., Sullivan, K. M., and Lilley, D. M. J.,** The structure of cruciforms in supercoiled DNA: probing the single-stranded character of nucleotide bases with bisulphite, *EMBO J.,* 5, 191, 1986.

177. **Lilley, D. M. J. and Dahlberg, J., Eds.,** *Nonstandard DNA Strutures and their Analysis, Methods in Enzymology,* Academic Press, Orlando, FL, 1992, 211.

178. **Voloshin, O. N., Mirkin, S. M., Lyamichev, V. I., Belotserkovskii, B. P., and Frank-Kamenetskii, M. D.,** Probing of irregular homopurine-homopyrimidin inserts by gel electrophoresis and diethyl pyrocarbonate, *Nature,* 333, 475, 1988.

179. **Htun, H. and Dahlberg, J. E.,** Single strands, triple strands, kinks in H-DNA, *Science* 241, 1791, 1988.

180. **Johnston, B. H.,** The S1-sensitive form of $d(CT)_n \cdot d(AG)_n$: chemical evidence for a three-stranded structure in plasmids, *Science* 241, 1800, 1988.

181. **Dean, W. W. and Lebowitz, J.,** Partial alteration of secondary structure in native superhelical DNA, *Nature, New Biol.,* 231, 5, 1971.

182. **Wang, J. C.,** Interaction between twisted DNAs and enzymes: the effects of superhelical turns, *J. Mol. Biol.,* 87, 797, 1974.

183. **Lebowitz, J., Garon, C. G., Chen, M. C. Y., and Salzman, N. P.,** Chemical modification of SV40 DNA by reaction with water-soluble carbodiimide, *J. Virol.,* 18, 205, 1976.

184. **Maxam, A. M. and Gilbert, W.,** A new method for sequencing DNA, *Proc. Natl. Acad. Sci. U.S.A.,* 74, 560, 1977.

185. **Lilley, D. M. J. and Hallam, L. R.,** Thermodynamics of the ColE1 cruciform. Comparisons between probing and topological experiments using single topoisomers, *J. Mol. Biol.,* 180, 179, 1984.

186. **Lyamichev, V. I., Voloshin, O. N., Frank-Kamenetskii, M. D., and Soyfer, V. N.,** Photofootprinting of DNA triplexes, *Nucl. Acids Res.,* 19, 1633, 1991.

187. **Courey, A. J. and Wang, J. C.,** Cruciform formation in a negatively supercoiled DNA may be kinetically forbidden under physiological conditions, *Cell,* 33, 817, 1983.

188. **Lafer, E. M., Moller, A., Nordheim, A., Stollar, B. D., and Rich, A.,** Antibodies specific for left-handed Z-DNA, *Proc. Natl. Acad. Sci. U.S.A.,* 78, 3546, 1981.

189. **Moller, A., Gabriels, J. E., Lafer, E. M., Nordheim, A., Rich, A., and Stollar, B. D.,** Monoclonal antibodies recognize different parts of Z-DNA, *J. Biol. Chem.,* 257, 12081, 1982.

190. **Malfoy, B. and Leng, M.,** Antiserum to Z-DNA, *FEBS Lett.,* 132, 45, 1981.

191. **Thomae, R., Beck, S., and Pohl, F. M.,** Isolation of Z-DNA containing plasmids, *Proc. Natl. Acad. Sci. U.S.A.,* 80, 5550, 1983.

192. **Di Capua, E., Stasiak, A., Koller, T., Brahms, S., Thomae, R., and Pohl, F. M.,** Torsional stress induces left-handed helical stretches in DNA of natural base sequence: circular dichroism and antibody binding, *EMBO J.,* 2, 1531, 1983.

193. **Revet, B., Zarling, D. A., Jovin, T. M., and Delain, E.,** Different Z-DNA forming sequences are revealed in ϕX174 RFI by high resolution darkfield immuno-electron microscopy, *EMBO J.,* 3, 3353, 1984.

194. **Hagen, F. K., Zarling, D. A., and Jovin, T. M.,** Electron microscopy of SV40 DNA cross-linked by anti-Z DNA IgG, *EMBO J.,* 4, 837, 1985.

195. **Lafer, E. M., Sousa, R., Ali, R., Rich, A., and Stollar, B. D.,** The effect of anti-Z-DNA antibodies on the B-DNA-Z-DNA equilibrium, *J. Biol. Chem.,* 261, 6438, 1986.

196. **Gellert, M., Mizuuchi, K., O'Dea, M. H., Ohmori, H., and Tomizawa, J.,** DNA gyrase and DNA supercoiling, *Cold Spring Harbor Symp. Quant. Biol.,* 43, 35, 1979.

197. **Wang, J. C., Peck, L. J., and Becherer, K.,** DNA supercoiling and its effects on DNA structure and function, *Cold Spring Harbor Symp. Quant. Biol.,* 47, 85, 1983.

198. **Lyamichev, V. I., Panyutin, I. G., and Frank-Kamenetskii, M. D.**, Evidence of cruciform structures in superhelical DNA provided by two-dimensional gel electrophoresis, *FEBS Lett.*, 153, 298, 1983.

199. **Panyutin, I., Lyamichev., V., and Mirkin, S.**, A structural transition in $d(AT)_n \cdot d(AT)_n$ inserts within superhelical DNA, *J. Biomol. Struct. Dyn.*, 2, 1221, 1985.

200. **Gough, G. W. and Lilley, D. M. J.**, DNA bending induced by cruciform formation, *Nature*, 313, 154, 1985.

201. **Kang, D. S. and Wells, R. D.**, B-Z junctions contain few, if any, nonpaired bases at physiological superhelical densities, *J. Biol. Chem.*, 260, 7783, 1985.

202. **Frank-Kamenetskii, M. D. and Vologodskii, A. V.**, Thermodynamics of the B-Z transition in superhelical DNA, *Nature*, 307, 481, 1984.

203. **Kozyavkin, S. A., Slesarev, A. I., Malkhosyan, S. R., and Panyutin, I. B.**, DNA linking potential generated by gyrase, *Eur. J. Biochem.*, 191, 105, 1990.

204. **Sanger, F., Coulson, A. R., Friedmann, T., Air., C. M., Barrell, B. G., Brown, N. L., Fiddes, J. C., Hutchison, C. A., III, Slocombe, P. M., and Smith, M.**, The nucleotide sequence of bacteriophage φX174, *J. Mol. Biol.*, 125, 225, 1978.

205. **Vologodskii, A. V., Lukashin, A. V., Anshelevich, V. V., and Frank-Kamenetskii, M. D.**, Fluctuations in superhelical DNA, *Nucl. Acids Res.*, 6, 967, 1979.

206. **Vologodskii, A. V. and Frank-Kamenetskii, M. D.**, Theoretical study of cruciform states in superhelical DNAs, *FEBS Lett.*, 143, 257, 1982.

207. **Singleton, C. K. and Wells, R. D.**, Relationship between superhelical density and cruciform formation in plasmid pVH51, *J. Biol. Chem.*, 257, 6292, 1982.

208. **Naylor, L. H., Lilley, D. M. J., and Van de Sande, J. H.**, Stress-induced cruciform formation in cloned $d(CATG)_{10}$ sequence, *EMBO J.*, 5, 2407, 1986.

209. **Naylor, L. H., Yee, H. A., and Van de Sande, J. H.**, Length-dependent cruciform extraction in $d(CATG)_n$ sequences, *J. Biol. Struct. Dyn.*, 5, 895, 1988.

210. **Greaves, D. R., Patient, R. K., and Lilley, D. M. J.**, Facile cruciform formation by an (A-T) sequence from a xenopus globin gene, *J. Mol. Biol.*, 185, 461, 1985.

211. **Singleton, C. K.**, Effects of salts, temperature, and stem length on supercoil-induced formation of cruciforms, *J. Biol. Chem.*, 258, 7661, 1983.

212. **Panyutin, I., Klishko, V., and Lyamichev, V.**, Kinetics of cruciform formation and stability of cruciform structure in superhelical DNA, *J. Biomol. Struct. Dyn.*, 1, 1311, 1984.

213. **Dayn, A., Malkhosyan, S., Duzhy, D., Lyamichev, V., Panchenko, Y., and Mirkin, S.,** Formation of (dA-dT)$_n$ cruciform in *E. coli* cells under differenct environmental conditions, *J. Bacteriol.,* 173, 2658, 1991.

214. **Iacono-Connors, L. and Kowalski, D.,** Altered DNA conformations in the regulatory region of torsionally-stressed SV40 DNA, *Nucl. Acids Res.,* 14, 8949, 1986.

215. **Brahms, S., Vergne, J., and Brahms, J. G.,** Natural DNA sequences form left-handed helices in low salt solution under conditions of topological constraint, *J. Mol. Biol.,* 162, 473, 1982.

216. **Pohl, F. M., Thomae, R., and Di Capua, E.,** Antibodies to Z-DNA interact with form V DNA, *Nature,* 300, 545, 1982.

217. **Haniford, D. B. and Pulleyblank, D. E.,** Facile transition of poly (d(TG)·d(CA)) into a left-handed helix in physiological conditions, *Nature,* 302, 632, 1983.

218. **Peck, L. J. and Wang, J. C.,** The energetics of B to Z transition in DNA, *Proc. Natl. Acad. Sci. U.S.A.,* 80, 6206, 1983.

219. **Stirdivant, S. M., Klysik, J., and Wells, R. D.,** Energetic and structural inter-relationship between DNA supercoiling and the right- to left-handed Z helix transitions in recombinant plasmids, *J. Biol. Chem.,* 257, 10159, 1982.

220. **Azorin, F., Nordheim, A., and Rich A.,** Formation of Z-DNA in negatively supercoiled plasmids is sensitive to small changes in salt concentration within the physiological range, *EMBO J.,* 2, 649, 1983.

221. **Pohl, F. M.,** Salt-induced transition between two double helical forms of oligo (dC-dG), *Cold Spring Harbor Symp. Quant. Biol.,* 47, 113, 1983.

222. **Frank-Kamenetskii, M. D., Lukashin, A. V., and Anshelevich, V. V.,** Application of polyelectrolyte theory to the study of B-Z transition in DNA, *J. Biomol. Struct. Dyn.,* 3, 35, 1985.

223. **Roy, K. B. and Miles, H. T.,** A thermally driven interconversion of B- and Z-DNA, *Biophys. Biochem. Res. Commun.,* 115, 100, 1983.

224. **O'Connor, T. R., Kang, D. S., and Wells, R. D.,** Thermodynamic parameters are sequence-dependent for the supercoil-induced B to Z transition in recombinant plasmids, *J. Biol. Chem.,* 261, 13302, 1986.

225. **Wang, A. H.-J., Gessner, R. V., van der Marel, G. A., van Boom., J. H., and Rich. A.,** Crystal structure of Z-DNA without an alternating purine-pyrimidine sequence, *Proc. Natl. Acad. Sci. U.S.A.,* 82, 3611, 1985.

226. **Vologodskii, A. V. and Frank-Kamenetskii, M. D.,** Left-handed Z form in superhelical DNA: a theoretical study, *J. Biomol. Sturct. Dyn.,* 1, 1325, 1984.

227. **Ellison, M. J., Kelleher, R. J., III, Wang, A. H.-J., Habener, J. F., and Rich, A.,** Sequence dependent energetics of the B-Z transition in supercoiled DNA containing non-alternating purine-pyrimidine sequences, *Proc. Natl. Acad. Sci. U.S.A.*, 82, 8320, 1985.

228. **Mirkin, S. M., Lyamichev, V. I., Kumarev, V. P., Kobzev, V. F., Nosikov, V. V., and Vologodskii, A. V.,** The energetics of the B-Z transition in DNA, *J. Biomol. Struct. Dyn.*, 5, 79, 1987.

229. **Ellison, M. J., Feigon, J., Kelleher, R. J. III, Wang, A. H.-J., Haabener, J. F., and Rich A.,** An assessment of the Z-DNA forming potential of alternating dA-dT stretches in supercoiled plasmids, *Biochemistry*, 25, 3648, 1986.

230. **McLean, M. J., Blano, J. A., Kilpatrick, M. W., and Wells, R. D.,** Consecutive AT pairs can adopt a left-handed DNA structure, *Proc. Natl. Acad. Sci. U.S.A.*, 83, 5884, 1986.

231. **Ho, P. S., Ellison, M. J., Quigley, G. J., and Rich, A.,** A computer aided thermodynamic approach fog predicting the formation of Z-DNA in naturally occurring sequences, *EMBO J.*, 5, 2737, 1986.

232. **Anshelevich, V. V., Vologodskii, A. V., and Frank-Kamenetskii, M. D.,** Theoretical study of formation of DNA noncanonical structures under superhelical stress, *J. Biomol. Struct. Dyn.*, 6, 247, 1988.

233. **Vinograd, J., Lebowitz, J., and Watson, R.,** Early and late helix-coil transitions in closed circular DNA. The number of superhelical turns in polyoma DNA, *J. Mol. Biol.*, 33, 173, 1968.

234. **Gagua, A. V., Belinstev, B. N., and Lyubchenko, Yu. L.,** Effect of base pairs stability on the melting of superhelical DNA, *Nature*, 294, 662, 1981.

235. **Melchior, W. B. and von Hippel, P. H.,** Alternation of the relative stability of dA·dT and dG·dC base pairs in DNA, *Proc. Natl. Acad. Sci. U.S.A.*, 70, 298, 1973.

236. **Voskoboinik, A. D., Monaselidze, D. R., Mgeladze, G. N., Chanchalashvili, Z. I., Lazurkin, Y. S., and Frank-Kamenetskii, M. D.,** Study of DNA melting in the region of the inversion of relative of stability of AT and GC pairs, *Mol. Biol.*, 9, 783, 1975.

237. **Laiken, N.,** Theoretical model for the equilibrium behavior of DNA superhelices, *Biopolymers*, 12, 11, 1973.

238. **Vologodskii, A. V. and Frank-Kamenetskii, M. D.,** Premelting of superhelical DNA: an expression for superhelical energy, *FEBS Lett.*, 131, 178, 1981.

239. **Burke, R. L. and Bauer, W. R.,** The early melting of closed duplex DNA: analysis by banding in buoyant neutral rubidium tri chloroacetate, *Nucl. Acids. Res.*, 8, 1145, 1980.

240. **Lee, F. S. and Bauer, W. R.,** Tenperature dependence of the gel electrophoretic mobility of superhelical DNA, *Nucl. Acids. Res.,* 13, 1665, 1985.

241. **Lyamichev, V. I., Mirkin, S. M., and Frank-Kamenetskii, M. D.,** A pH-dependent structural transition in the homoopurine-homopyrimidine tracts in superhelical DNA, *J. Biomol. Struct. Dyn.,* 3, 327, 1985.

242. **Lyamichev, V. I., Mirkin, S. M., and Frank-Kamenetskii, M. D.,** Structures of homopurine-homopyrimidine tracts in superhelical DNA, *J. Biomol. Struct. Dyn.,* 3, 667, 1986.

243. **Lyamichev, V. I., Mirkin, S. M., and Frank-Kamenetskii, M. D.,** Structure of $(dG)_n \cdot (dC)_n$ under superhelical stress and acid pH, *J. Biomol. Struct. Dyn.,* 5, 275, 1987.

244. **Mirkin, S. M., Lyamichev, V. I., Drushlyak, K. N., Dobrynin, V. N., Filippov, S. A., and Frank-Kamenetskii, M. D.,** The H form requires a homopurine-homopyrimidine mirror repeat, *Nature,* 330, 495, 1987.

245. **Htun, H. and Dahlberg, J. H.,** Topology and formation of triple-stranded H-DNA, *Science,* 243, 1571, 1989.

246. **Lyamichev, V. I., Mirkin, S. M., Kumarev, V. P., Baranova, L. V., Vologodskii, A. V., and Frank-Kamenetskii, M. D.,** Energetics of the B-H transition in supercoiled DNA carring $d(CT)_x \cdot d(AG)_x$ and $d(C)_n \cdot d(G)_n$ inserts, *Nucl. Acids Res.,* 17, 9417, 1989.

247. **Fox, K. R.,** Long $d(A) \cdot d(T)$ tracts can form intramolecular triplexes under superhelical stress, *Nucl. Acids Res.,* 18, 5387, 1990.

248. **Benham, C. J.,** Statistical mechanical analysis of competing conformational transitions in superhelical DNA, *Cold Spring Harbor Symp. Quant. Biol.,* 47, 219, 1983.

249. **Luchnik, A. N. and Glazer, V. M.,** Decrease in the number of DNA topological turns during friends erythroleukemia differentiation, *Mol. Gen. Genet.,* 178, 459, 1980.

250. **Krylov, D. Yu., Makarov, V. L., and Ivanov, V. I.,** B-A transition in superhelical DNA, *Nucl. Acids Res.,* 18, 759, 1990.

251. **Pakhomov, D. V. and Vologodskii, A. V.,** Theoretical analysis of B-A transition in circular DNA, *Biopolym. Cell,* 2, 240, 1986.

252. **Nordheim, A., Lafer, E. M., Peck, L. J., Wang, J. C., Stollar, B. D., and Rich, A.,** Negatively supercoiled plasmids contain left-handed Z-DNA segments as detected by specific antibody binding, *Cell,* 31, 309, 1982.

253. **Singleton, C. K., Klysik, J., and Wells, R. D.,** Conformation flexibility of junctions between contiguous B- and Z-DNAs in supercoiled plasmids, *Proc. Natl. Acad. Sci. U.S.A.,* 80, 2447, 1983.

254. **Vologodskii, A. V.,** Formation of noncanonical structures in superhelical DNA. Effects of mutual influence of transition, *Mol. Biol.*, 19, 568, 1985.

255. **Ellison, M. J., Fenton, M. J., Ho, P. S., and Rich, A.,** Long-range interactions of multiple of DNA structural transitions within a common topological domain, *EMBO J.*, 6, 1513, 1987.

256. **Kelleher, R. J., III, Ellison, M. J., Ho, P. S., and Rich, A.,** Competitive behavior of multiple, discrete B-Z transitions in supercoiled DNA, *Proc. Natl. Acad. Sci. U.S.A.*, 83, 6342, 1986.

257. **Mizuuchi, K., Muzuuchi, M., and Gellert, M.,** Cruciform structures in palindromic DNA are favored by DNA supercoiling, *J. Mol. Biol.*, 156, 229, 1982.

258. **Peck, L. J., Wang, J. C., Nordheim, A., and Rich, A.,** Rate of B to Z structural transition of supercoiled DNA, *J. Mol. Biol.*, 190, 125, 1986.

259. **Pohl, F. M.,** Dynamics of the B-to-Z transition in supercoiled DNA, *Proc. Natl. Acad. Sci. U.S.A.*, 83, 4983, 1986.

260. **Vologodskii, A. V. and Frank-Kamenetskii, M. D.,** The relaxation time for a cruciform structure in superhelical DNA, *FEBS Lett.*, 160, 173, 1983.

261. **Courey, A. J. and Wang, J. C.,** Influence of DNA sequence and supercoiling on the process of cruciform formation, *J. Mol. Biol.*, 202, 35, 1988.

262. **Gellert, M., O'Dea, M. H., and Mizuuchi, K.,** Slow cruciform transitions in palindromic DNA, *Proc. Natl. Acad. Sci. U.S.A.*, 80, 5545, 1983.

263. **Thompson, B. J., Camin, M. N., and Warner, R. C.,** Kinetics of branch migration in double-stranded DNA, *Proc. Natl. Acad. Sci. U.S.A.*, 73, 2299, 1976.

264. **Sinden, R. R. and Pettijohn, D. E.,** Cruciform transition in DNA, *Biol. Chem.*, 259, 6593, 1984.

265. **Sullivan, K. M. and Lilley, D. M. J.,** A dominant influence of flanking sequences on local structural transition in DNA, *Cell*, 47, 817, 1986.

266. **Wang, J. C.,** Helical repeat of DNA in solution, *Proc. Natl. Acad. Sci. U.S.A.*, 76, 200 1979.

267. **Peck, L. J. and Wang, J. C.,** Sequence dependence of the helical repeat of DNA in solution, *Nature*, 292, 375, 1981.

268. **Strauss, F., Gaillard, C., and Prunell, A.,** Helical periodicity of DNA, poly(dA)·poly(dT) and poly(dA-dT)·poly(dA-dT) in solution, *Eur. J. Biochem.*, 118, 215, 1981.

269. **Goulet, I., Zivanovic, Y., and Prunell, A.,** Helical repeat of DNA in solution. The V curve method, *Nucl. Acids Res.,* 15, 2803, 1987.

270. **Wang, J. C., Jacobsen, J. H., and Saucier, J.-M.,** Physicochemical studies of interactions between DNA and RNA polymerase. Unwinding of the DNA helix by *E. coli* RNA polymerase, *Nucl. Acids Res.,* 4, 1225, 1977.

271. **Richmond, T. J., Finch. J. T., Rushton, B., Rhodes, D., and Klug, A.,** Structure of the nucleosome core particle at 7 Å resolution, *Nature,* 3211, 532, 1984.

272. **Fuller, F. B.,** Decomposition of the linking number of a closed ribbon: a problem from molecular biology, *Proc. Natl. Acad. Sci. U.S.A.,* 75, 3557, 1978.

273. **Sinden, R. R., Carlson, J. O., and Pettijohn, D. E.,** Torsional tension in the DNA double helix measured with trimethylpsoralen in living *E. coli* cells: analogous measurements in insects and human cells, *Cell,* 21, 773, 1980.

274. **Bliska, J. B. and Cozzarelli, N. R.,** Use of site-specific recombination as a probe of DNA structure and metabolism in vivo, *J. Mol. Biol.,* 194, 205, 1987.

275. **Lyamichev, V., Panyutin, I., and Mirkin, S.,** The absence of cruciform structures from pAO3 plasmid DNA in vivo, *J. Biomol. Struct. Dyn.,* 2, 291, 1984.

276. **Sinden, R. R., Broyles, S. S., and Pettijohn, D. E.,** Perfect palindromic *lac* operator DNA sequence exists as a stable cruciform structure in supercoiled DNA in vitro but not in vivo, *Proc. Natl. Acad. Sci. U.S.A.,* 80, 1797, 1983.

277. **Borst, P., Overdulve, J. P., Weijers, P. J., Fase-Fowler, F., and Van der Berg, M.,** DNA circles with cruciforms from *Isospora* (Toxoplasma) *gondii, Biochem. Biophys. Acta,* 781, 100, 1984.

278. **Haniford, D. B. and Pulleyblank, D. E.,** The in vivo occurrence of Z DNA, *J. Biomol. Struct. Dyn.,* 1, 593, 1983.

279. **Pruss, G. J. and Drlica, K.,** Topoisomerase 1 mutants: the gene on pBR322 that encodes resistance to tetracycline affects plasmid DNA supercoiling, *Proc Natl. Acad. Sci. U.S.A.,* 83, 8952, 1986.

280. **Haniford, D. B. and Pulleyblank, D. E.,** Transition of a cloned $d(AT)_n \cdot d(AT)_n$ tract to a cruciform in vivo, *Nucl. Acids Res.,* 13, 4343, 1985.

281. **McClellan, J. A., Boublikova, P., Palecek, E., and Lilley, D. M. J.,** Superhelical torsion in cellular DNA responds directly to environmental and genetic factors, *Proc. Natl. Acad. Sci. U.S.A.,* 87, 8373, 1990.

282. **Zacharias, W., Jaworski, A., Larson, J. E., and Wells, R. D.,** The B- to Z-DNA equilibrium in vivo is perturbed by biological processes, *Proc. Natl. Acad. Sci. U.S.A.,* 85, 7069, 1988.
283. **Lui, L. F. and Wang, J. C.,** Supercoiling of DNA template during transcription, *Proc. Natl. Acad. Sci. U.S.A.,* 84, 7024, 1987.
284. **Wu, H.-Y., Shyy, S. S., Wang, J. C., and Liu, L. F.,** Transcription generates positively and negatively supercoiled domains in the template, *Cell,* 53, 433, 1988.
285. **Giaever, G. N. and Wang, J. C.,** Supercoiling of intracellular DNA can occur in eucaryotic cells, *Cell,* 55, 849, 1988.
286. **Droge, P. and Nordheim, A.,** Transcription-induced conformational change in a topologically DNA domain, *Nucl. Acids Res.,* 19, 2941, 1991.
287. **Wang, J. C. and Liu, L. F.,** DNA replication: topological aspects and the roles of DNA topoisomerases, in *DNA Topology and Its Biological Effects,* Cozzarelli, N. R. and Wang, J. C., Eds., Cold Spring Harbor Laboratory Press, Cold Spring Harbor, NY, 1990, 321.
288. **Rahmouni, A. R. and Wells, R. D.,** Stabilization of DNA in vivo by localized supercoiling, *Science,* 246, 358, 1989.

INDEX